# *Farming System Approaches for Sustainable Hill Agriculture in Manipur*

# *Farming System Approaches for Sustainable Hill Agriculture in Manipur*

*—Authors—*

**N. Prakash**

**M.A. Ansari**

**S.S. Roy**

**P. Punitha**

**L.K. Baishya**

**Blessa Sailo**

**S.K. Sharma**

**S.V. Ngachan**

*ICAR Research Complex for NEH Region*
*Manipur Centre, Lamphelpat, Imphal*

BIOTECH

2017

BIOTECH BOOKS®

ISBN 978-81-7622-396-6



**Correct Citation :** Prakash N, Ansari M A, Roy S S, Punitha P, Baishya L K, Sailo Blessa, Sharma S K and Ngachan S V. 2017. Farming System Approaches for Sustainable Hill Agriculture in Manipur, 185 pages. Biotech Books, New Delhi.

*Published by*: **BIOTECH BOOKS®**
4762-63/23, Ansari Road, Darya Ganj,
New Delhi - 110 002
Phone: +91-011-23262132
E-mail: biotechbooks@yahoo.co.in

*Typeset at*: **Classic Computer Services**
Delhi - 110 035

*Digitally Printed at:* **Replika Press Pvt. Ltd.**

PRINTED IN INDIA

क. अलगुसुन्दरम
उप महानिदेशक (अभियांत्रिकी)
**K. Alagusundaram**
Deputy Director General (Engineering)

भा.कृ.अ.प.
ICAR

**भारतीय कृषि अनुसंधान परिषद**
कृषि अनुसंधान भवन–II,
पूसा, नई दिल्ली 110 012
**INDIAN COUNCIL OF AGRICULTURAL RESEARCH**
KRISHI ANUSANDHAN BHAVAN-II,
PUSA, NEW DELHI-110 012

# Foreword

Technologies that tend to homogenize the natural environment and agricultural production have low possibilities of success in environmental stress conditions that are likely to result from climate change. On the other hand, technologies that allow for, and promote diversity are more likely to provide a strategy which strengthens agricultural production in the face of uncertain future climate change scenarios. Technologies and practices do exist, or have been developed in different parts of our country, to facilitate adaptation to climate change in the agriculture sector. Integrated Farming System has emerged as one of the important facets of climate smart agriculture for a climate compatible development particularly in the context of fragile hill ecosystem. Integrated Farming System is highly knowledge intensive, location specific and its success depends on the choice of enterprise, farmers' capability to manage the different enterprises, scientific integration of various farm components and socio-economic situation. Hence, it is of utmost importance to document the findings of various works undertaken on integrated farming system and bring those findings in the public domain. IFS should be popularized not only to ensure food and nutritional security, but should also be looked into bio-physical or environmental perspective.

I am immensely happy that ICAR Research Complex for NEH Region is bringing out this book entitled "Farming System Approaches for Sustainable Hill Agriculture in Manipur" which I feel will cater a long felt need of the researchers, academicians, policy makers, students and most importantly farmers. This book is a outcome of participatory research and extension activities undertaken in different hill agro-ecosystem of Manipur by ICAR which also includes resource inventory, IFS modules, case studies and success stories.

Ph: 91-11-2584 3415 Fax: 91-11-2584 2660 E-mail: ddgengg@icar.org.in

I compliment the ICAR Research Complex for NEH Region, Manipur Centre for coming out with this publication. I am sure this publication will be an important resource for all the stakeholders in agriculture.

29/11/16

*K. Alagusundaram*

# Preface

Excellent and prime agricultural technologies are of limited value and scope if they are not adopted and accepted by the farmers. The farmers need those technologies which increase their net returns and livelihood with keeping minimum risk and within their own purview and management practices. In the era of modernization, the traditional transfer of technology models are inadequate to meet the multiple objectives of poverty reduction, food security, competitiveness, sustainability and demand of the farming community. Further, to meet the specific client demand and to transfer location specific high income generating integrating farming system models, participatory research is emphasised in India in general and North East India in particular. Attempts have been made to integrate the desirable components of farming system research into the mainstream agricultural research so that the technologies developed are relevant, client-oriented and location specific. The Agriculture in the Manipur is complex, diverse and risk prone (CDR). Almost 80 per cent population of the region is dependent on agriculture. The farmers are resource poor and practice agriculture mostly for subsistence farming. About 85 per cent of the farmers are small and marginal but share only 40 per cent of the land. The hill soil is fragile and less fertile with high acidic in reaction. The crop productivity is very low mainly due to lack of resources, non-adoption of improved technologies and fragile ecosystems. Integrated farming system is a reliable way of obtaining high productivity with substantial nutrient economy for higher resource use efficiency in combination with maximum compatibility and replenishment of organic matter by way of effective recycling of organic residues/wastes etc. obtained through integration of various land- based enterprises. The modules mentioned in this book under transfer of technology and research activities organized by ICAR Research Complex for NEH region, Manipur Centre, Imphal is to impart knowledge about the methodology, benefits and component selection at the farmer's field to increase the livelihood of the resource poor farmers with limited land and other resources. In this book, we share the experiences related to integrated farming system at farmers field from enterprise selection to integration of various enterprises for improving the livelihood of the farmers in the hilly area of Manipur. We hope that the book will be a valuable source of information for students, teachers, scientists, extension functionaries and policy planners.

***Authors***

# Contents

# The Authors

**Dr. Narendra Prakash** did his Ph.D. in Agricultural Extension from Meerut University. He was the Head, Division of Agricultural Extension at ICAR Research Complex for NEH Region, Umiam, Meghalaya from 1997-2007. Presently, he is Joint Director, ICAR Research Complex for NEH Region, Manipur Centre, Imphal since 2007. Dr. Prakash has vast experience on hill agriculture, indigenous technical knowledge and has developed various farming system models in North Eastern Hill region. He is very much instrumental in organizing capacity building programme, demonstration and agripreneurship development for greater benefit of the farming community in the region.

Under leadership of Dr. Prakash, eight high yielding varieties of rice, tomato, Jatropha etc. to has been developed from ICAR Research Complex for NEH Region, Manipur Centre, Imphal and standardized agro-technologies. Dr. Prakash has authored/edited 10 books and published more than 50 research papers, 16 book chapters and many popular articles, technical bulletins and manuals on hill agriculture.

**Dr. Meraj Alam Ansari** (ARS) is an Alumni of ICAR- Indian Agriculture Research Institute, New Delhi and was outstanding in academic career. He is recipient of Chancellor and Vice Chancellor Gold Medal in M. Sc. Ag. (Agronomy) at NDUAT, Kumarganj, Faizabad. Presently he is serving as a Scientist (Agronomy) at ICAR Research Complex for NEH Region, Manipur Centre, Imphal since 27th April, 2011. During this period, he has developed farmer's friendly, economically viable and environmentally sustainable resource conserving technologies (RCT) in cereals and pulses, integrated farming system (IFS), biomass management etc. for ensuring food and nutritional security in North Eastern Hill region. He is also engaged in climate change research and working on the potential of C- sequestration in various land use system in hilly terrains. He is instrumental in disseminating various technologies to the farmer's field. He has published 32 research papers in national and international repute journals, 7 book chapters, 2 review papers, 2 conference books of abstracts, 23 popular articles and souvenir, 7 training and technology bulletin etc. For his significant contribution in agricultural research and extension, he has received many prestigious awards.

**Dr. S.S. Roy** joined the Agricultural Research Service in 2008. He obtained his Doctoral Degree in Spices and Plantation Crops from Bidhan Chandra Krishi Viswavidyalaya, West Bengal. Dr. S. S. Roy, has an excellent academic career and is recipient of University Gold Medal and Certificate of Merit. He is serving as Scientist (Horticulture) at ICAR Research Complex for NEH Region, Manipur Centre for more than eight years in conceptualizing and developing technologies for enhancing productivity and profitability of horticultural crops. Dr. Roy is actively engaged in popularization of integrated farming system and organic crop production in hilly region. In addition to three high yielding crop varieties to his credit; he has also standardized more than twenty agro-technologies. He was one of the team members for standardization of DUS-Testing guidelines for Khasi mandarin. He was a visiting scholar at Michigan State University, USA under NAIP International Training Programme. Dr. Roy has published more than 30 research papers in the journals of national and international repute, 4 books, 14 book chapters, 43 popular articles, 8 bulletins and 6 manuals. He is the editorial board member of three national journals. Dr. Roy has received many prestigious awards. His findings have been widely adopted by the stakeholders and brought substantial development in productivity and livelihood of the resource poor farmers of North Eastern Hill Region of India.

**Mrs. Punitha. P** (ARS) is serving as a Scientist (Agricultural Extension) at ICAR Research Complex for NEH Region, Manipur Centre, Imphal. Presently she is doing her Ph. D. (Agricultural Extension) on livelihood studies of Jhumias in Manipur at College of P.G. studies, CAU, Umiam, Meghalaya. To her credit, she has done research on Communication, Impact Assessment and Integrated Farming System. She has accomplished many Front Line demonstrations (FLDs), farmer trainings, Field days on Rice, Lentil, Integrated farming system and Groundnut for the benefit of both men, women and youth farmers in coordination with Krishi Vigyan Kendra of ICAR in Manipur Region. She has published 7 research papers, 1 review paper, 8 technical/ Extension bulletin, 3 book chapters, 3 popular articles. She has received two best papers, a best poster and a research scholar award for her research and extension.

**Dr. Lohit K. Baishya** is working as a Senior Scientist (Agronomy), ICAR Research Complex for NEH Region, Manipur Centre. His specialized area of research is soil fertility, crop nutrition & Crop diversification. Now he is working on Jhum (Shifting cultivation) improvement, National Innovations on Climate Resilient Agriculture, National Mission on Sustaining Himalayan Ecosystem and Tribal sub plan. He has published more than 30 papers of national and international repute.

**Dr. Blessa Sailo** is an ICAR-Indian Veterinary Research Institute, Izzatnagar, Bareilly, Uttar Pradesh alumni with specialization in Veterinary Public Health. Presently he is working as Scientist (Veterinary Public Health) at ICAR Research Complex for NEH Region, Manipur Centre, Imphal since 2012. His research priorities include characterization of Multi-Drug resistant Enterobacteriaceae from food producing animals as well as foods of animal origin in North Eastern Hill (NEH) region of India and surveillance of economically important livestock's diseases. He has worked on scientific management of livestock's with reference to NEH region and has been instrumental in dissemination of improved germplasm of poultry birds and pigs to the farmer fields. He has worked on integration of livestock's in intensive integrated farming systems (IIFS) being developed and popularized by ICAR Research Complex for NEH Region, Manipur Centre.

**Dr. Susheel Kumar Sharma** is an ICAR-Indian Agriculture Research Institute, New Delhi alumni with specialization in Plant Pathology. Presently he is working as Scientist (Plant Pathology) at ICAR Research Complex for NEH Region, Manipur Centre, Imphal since 2013. His research priorities include characterization of emerging and reemerging plant diseases in North Eastern Hill (NEH) region of India, development of specific diagnostics and eco-friendly integrated disease management modules. He has worked on standardization of production practices of various mushroom species with reference to NEH region and has been instrumental in dissemination mushroom enterprises to the farmer fields. He has worked on integration of mushroom in intensive integrated farming systems (IIFS) being developed and popularized by ICAR Research Complex for NEH Region, Manipur Centre. He has published 12 research papers, 6 book chapters, 15 popular articles, 2 conference books of abstracts and souvenir, 3 training manuals, 2 leaflets and folder etc. He is the recipient of many academic awards for his significant contribution in the field of plant disease research.

**Dr. S.V. Ngachan** did his doctoral degree programme in Plant Pathology from Moscow State University, Russia. He was the Joint Director of Nagaland Centre and Manipur Centre of ICAR Research Complex for NEH Region, and at present, he is Director of the ICAR Research Complex for NEH Region, Umiam, Meghalaya. Dr. Ngachan has authored many books on hill agriculture and published more than 100 research papers and many popular articles in different national and international journals. He has vast experience in agricultural and socio-economic scenario of North Eastern Hill Region.

# 1

# Introduction

The major problems of present day agriculture are decline in agriculture growth rate, decline in factor productivity, static or decline in food production, increasing malnutrition, shrinkage in net cultivable area, increasing environmental pollution, depleting groundwater table, increasing cost of production, low farm income and increasing unemployment. The human population of India has increased to 1210.2 million at a growth rate of 1.76 per cent in 2011 and is estimated to increase further to 1530 million by 2030. There are projections that demand for food grains would increase from 250 million tonnes to 345 million tonnes in 2030. The Indian economy is predominantly rural and agricultural, and the declining trend in size of land holding poses a serious challenge to the sustainability and profitability of farming. In view of the decline in per capita availability of land from 0.5 ha in 1950-51 to 0.15 ha by the turn of the century and a projected further decline to less than 0.1 ha by 2020, it is imperative to develop strategies and agricultural technologies that enable adequate employment and income generation, especially for small and marginal farmers who constitute more than 80 per cent of the farming community. Declining size of landholdings without any alternative income augmenting opportunity is resulting in fall of farm income and causing agrarian distress.

To meet the multiple objectives of poverty reduction, food security, competitiveness and sustainability, several researchers have recommended the farming systems approach to research and development. A farming system is the result of complex interactions among a number of inter-dependent components, where an individual farmer allocates certain quantities and qualities of four factors of production, namely land, labour, capital and management to which he has access (Mahapatra, 1994). This is a multidisciplinary whole-farm approach and very effective in solving the problems of small and marginal farmers. The approach aims at increasing income and employment from small-holdings by integrating various farm enterprises and recycling crop residues and by-products within the farm itself (Behera and Mahapatra, 1999; Singh *et al.*, 2006). No single farm enterprise is likely to be able to sustain the small and marginal farmers without resorting to integrated farming systems (IFS) for the generation of adequate income and gainful employment year round (Mahapatra, 1992; 1994). Farming systems approach, therefore, is a valuable approach to addressing the problems of sustainable economic growth for farming communities in India. The basic aim of integrated farming system (IFS) is to derive a set of resource development and utilization practices, which lead to substantial and sustained increase in agricultural production (Kumar and Jain, 2005). Integrated farming systems are often less risky, if managed efficiently, they benefit from synergisms among enterprises, diversity in produce, and environmental soundness (Lightfoot, 1990). On this basis, IFS models have been suggested by several workers for the development of small and marginal farms across the country (Rangaswamy *et al.*, 1996; Behera and Mahapatra, 1999; Singh *et al.*, 2006).

## Farming System (FS)

Farming is a process of harnessing solar energy in the form of economic plant and animal products; whereas, system implies a set of interrelated process organized into a functional entity. Hence, a farming system is defined as a population of individual farm systems that have broadly similar resource bases, enterprise patterns, household livelihoods and constraints, and for which similar development strategies and interventions would be appropriate. Depending on the scale of the analysis, a farming system can encompass a few dozen or many millions of households. Farming systems of a region are decided by and large, by a number of soil and climatic parameters which determine overall agro-ecological setting for nourishment and appropriateness of crops and livestock or set of agriculture enterprise. At farmers' level, potential productivity and monetary benefits act as guiding principles while opting for a particular farming system. The decisions *w.r.t.* choice of farming systems are further narrowed down under influence of several other forces related to infrastructure facilities, socio-economic factors and technological developments, all operating interactively at micro-level. The household, its resources and the resource flows and interactions at the individual farm levels are together referred to as a farm system (FAO, 2001). Systems could be defined as an organised unitary whole composed of two or more inter dependant and interacting parts, components or subsystems delineated by identifiable boundary or its environmental super system. It is a set of interrelated elements each of which is associated directly or indirectly with other elements and no subset is under-related to any other subsets. In system approach, the whole farms rather than the individual crops/enterprises is considered before any decision related to the choice of enterprise and or technology is made. The farming systems can also be described and understood as by its structure and functioning. The structure in its wider sense includes among others, the land use pattern, production relations, land tenures, size of holding and their distribution, irrigation, marketing including transport and storage, credit institutions and financial markets and research and education.

Farming system designates a set of agricultural activities organized into functional units:

- ☆ To profitably harness solar energy
- ☆ Preserving land productivity
- ☆ Ensuring environmental quality
- ☆ Maintaining desirable level of biological diversity and ecological stability.

Resource Management strategies achieving economy and sustaining agriculture production

- ☆ Meeting diverse requirements of farming households
- ☆ Conserving the resource base and maintaining environment quality
- ☆ Efficient use of land, labour and available resources

Therefore, farming system is a complex inter-related matrix of soil, plants, animals, implements, power, labour, capital and other inputs controlled in parts by farming families and influenced to varying degrees by political, economic, institutional and social forces that operate at many levels (Mahapatra, 1992).

Factors determining type of farming:

- ☆ Physical factor (Climate, soil, topography)
- ☆ Economic factor (Marketing cost, labour availability, capital, land value, competition for enterprises, consumer demand, prevalent pest and diseases)
- ☆ Social factor (Type of community, easy transport, marketing facilities and co-operative spirit)

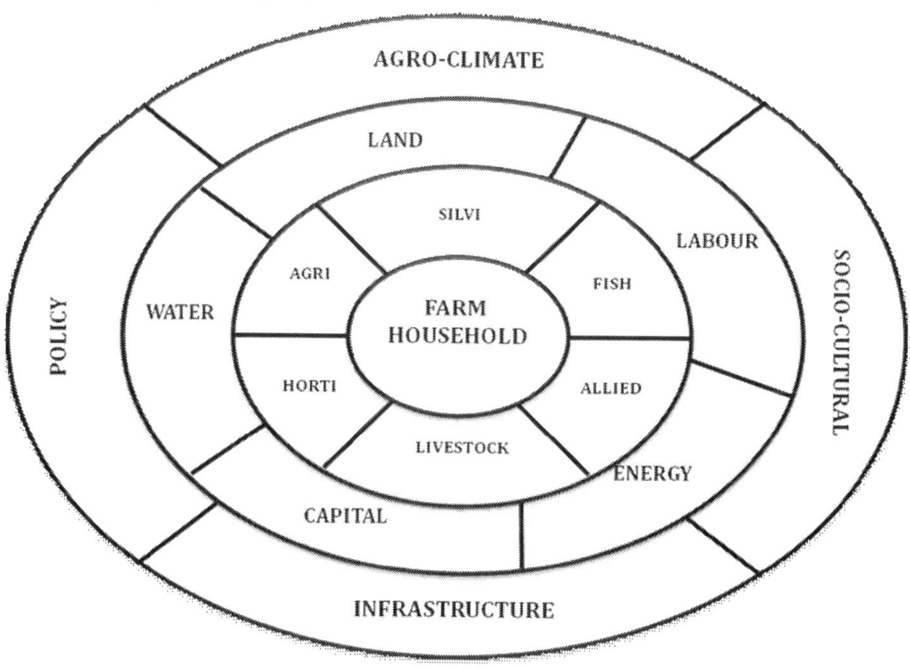

- ☆ Objective (Income, production, minimizing cost etc.)
- ☆ Availability of resources and components

Farming system focuses on the interdependencies between components under the control of household and how these components interact with the physical, biological and socio-economic factors, which is not under the control of household. Farm household is the basic unit of farming system and interdependent farming enterprises carried out on the farm. Farmers are subjected to many socio-economic, bio-physical, institutional, administrative and technological constraints. The operator of the farming system is farmer or the farming family. The highly simplified model of farming system puts the farmer the decision maker, at the centre. Decisions are influenced by the priorities of the household, farmer's knowledge and experiences, and resource at his command. External factors - natural, economic and sociocultural, also plays significant roles.

## Rainfed Farming Systems (RFS)

Rainfed agriculture occupies 68 per cent of India's cultivated area and supports 40 per cent of the human and 60 per cent of the livestock population. It produces 445 of the food requirement, thus has and will continue to play a critical role in India's food security. Agriculture in the rainfed areas and fragile ecosystems is inevitable for meeting the food, fibre and energy needs of the local inhabitants. If the rainfall is adequate, farmers can expect a bumper crop. In case of less rainfall, crop may fail and it may be lost completely if rainfall is low (drought). On the other side, a heavy rainfall may results in uncertainty in crop production and it may lose or survive depending on the severity of flood. However, aberrant behaviour of monsoon rainfall, eroded and degraded soils with multiple nutrient and water deficiencies, declining ground water table and poor resource base of the farmers are principle constraints for low and unstable yields in rainfed areas (Singh *et al.*, 2005). The guiding principles for selection of crops and varieties for efficient management of resources in rainfed areas are land use capability concept, water availability concept, crop substitution, quantity and distribution of rainfall, soil depth and performance of crops. Effective growing period concept is mostly used in deciding cropping systems in different agro-climatic zones. Traditionally rainfed farmers are small subsistent land holders integrating livestock with crop production. With continuing population growth, intensifying crop and livestock systems continue to play vital role in maintaining rural livelihoods. The philosophy behind shifting from cropping system to the farming system mode involves (i) *in situ* recycling of organic residues including farm wastes generated at the farm to reduce the dependency on chemicals (ii) decrease in cost of cultivation through enhance input use efficiency, (iii) effective use of by-products/ wastes of one component for the benefit of other component/components (iv) upgrading of soil and water quality and bio-diversity, (v) increased water productivity, (vi) nutritional security through minimizing chemical residues in soil plant animal human chain, and (vii) environmental security by moderating flow of green house gases from the soil to environment. Farming system provides a vast canvass of livelihood gathering, a better risk coping strategy, continuous flow of income and employment throughout the year for small landholders. It involves utilization of primary and secondary produce of one system as a basic input of other system through making them mutually integrated. In kharif season, the rainfall both in terms of quantum and distribution decides the effective growing season and it becomes critical in selecting cropping systems for a given reason. In rabi season available moisture in soil profile at sowing time dictates the choice of crops in a given cropping system. In regions, receiving 350-600 mm of rainfall and 20 weeks effective growing season, intercropping (150 per cent cropping intensity) is possible in regions having 20-30 weeks of effective growing season from 650-750 mm of rainfall. In areas receiving more than 750 mm of rainfall and having an effective growing season of more than 30 weeks double cropping is assured (Singh and Subba Reddy, 1986). Early planting and harvesting at physiological maturity of crops, less number of tillage operations, deep placement of fertilizers for rabi crops are crucial for succeeding double cropping.

## Farming Systems Research (FSR)

The farming system research activities are to be farmer oriented, system oriented, problem solving approach, inter-disciplinary, compliments mainstream disciplinary research, test the technology in on-farm trials and provides feedback from the farmers. The strategy of FSR should emphasize that the research agenda should be determined by explicitly define farmers needs through an understanding of the existing farming systems rather than its perception by the researchers. The farming systems research and extension should be dealt in holistic manner on farmers participatory mode with problem solving approach, keeping genders activity, interdisciplinary and interactive approach. It should emphasize extensive on-farm activities and complement the experimental on-station research and acknowledges the location specificity of technical solutions and document the inter dependencies among multiple clients. Greater importance is placed on feedback to modify the content of subsequent on farm trials, if necessary, by changing research priorities focusing policy shifts based upon micro level analysis. Hence farming systems research is designed to understand farmers' priorities, strategies and resource allocation decisions.

The core characteristics of the farming system research can be describes as follows.

- ☆ **It is problem solving:** As an applied problem solving approach, it emphasizes on developing and transferring appropriate technologies to overcome production constraints through diagnosis of biophysical, socio-economic and institutional constraints that influence technological solutions.
- ☆ **It is holistic:** The whole farm is viewed as a system encompassing interacting sub-systems, and no potential enterprise is considered in isolation.
- ☆ **It acknowledges the location specificity of technological solutions:** Recognizing the location specific nature of agricultural production problems, it emphasizes on testing and adaptation of technological solutions based on agro-ecological and socio-economic specificities.
- ☆ **It defines specific client groups:** Emphasis is made on the identification of specific and relatively homogeneous groups of farmers with similar problems and circumstances for whom technology is to be developed as

the specific client groups. On the basis of common environmental parameters, production patterns and management practices, relatively homogeneous recommendation domains need to be identified.

- **It is farmer participatory:** It revolves round the basic principle that successful agricultural research and development efforts should start and end with the farmers (Rhoades and Booth, 1982). Farmer participation is ensured at different stages of technology generation and transfer processes such as system description, problem diagnosis, design and implementation of on-farm trials, and providing feedback through monitoring and evaluation.
- **It gives weightage to ITK system:** The Indigenous Technical Knowledge (ITK), which is time tested at the farmer's level for sustainability through a dynamic process of integrating new innovations into the system as they arise, has to be properly understood by the scientists and utilised in their research activities.
- **It is concerned with 'Bottom-up' research strategy:** It begins with an understanding of existing farming system and the identification of key production constraints.
- **It is interdisciplinary:** It lays greater emphasis on interdisciplinary cooperation among the scientists from different areas of specialisation to solve agricultural problems that are of concern to farmers.
- **It emphasizes extensive on-farm activities:** It involves problem analysis through diagnostic surveys, on-farm testing of the developed technologies, and providing feedback through evaluation to influence the research agenda of the experiment stations. It provides a structural framework for the farmers to express their preferences and apply their evaluation criteria for selecting technologies suiting to their circumstances.
- **It is gender sensitive:** While explicitly acknowledging the gender-differentiated roles of farm family in agriculture, it emphasizes the critical review of farming systems in terms of activities analysis, access and control over resources and benefits and implication's in developing relevant research agenda.
- **It is iterative:** Instead of trying to know everything about a system at a time, it requires step-by-step analysis of only key functional relationships.
- **It is dynamic:** It involves recurrent analysis of the farming systems, permitting continuous learning and adaptations.
- **It recognizes interdependencies among multiple clients:** The generation, dissemination and adoption of relevant technologies to improve the productivity and sustainability of agriculture require productive and interactive linkages among the policy planners, scientists, developmental agencies and farmers. The approach attaches more importance for this critical factor.
- **It focuses on actual adoption:** It is to be judged by the extent to which it influences the production of socially desirable technologies that diffuse quickly amongst specified groups of farmer clients.

- ☆ **It focuses on sustainability:** It seeks to harness the strengths of the existing farming practices, and to ensure that productivity gains are environmentally acceptable. Towards preserving the natural resource base and strengthening the agricultural production base, it attempts to develop technologies that are environment friendly and economically viable.
- ☆ **It complements experiment station research:** It only complements but does not substitute on station research. It has to draw upon the scientific knowledge and technologies generated at research stations. It has to be kept in mind that the approach is not being promoted as panacea for all maladies of local agricultural production systems.

## Integrated Farming System (IFS)

Integrated farming system is the scientific integration of different interdependent and interacting farm enterprises for the efficient use of land, labour and other resources of a farm family which provide year round income to the farmers specially located in the handicapped zone. An IF a component of farming system research (FSR), introduces a change in farming techniques for maximum production is a cropping pattern and take care of optimal utilization of resources. Unlike specialized farming system (SFS) integrated farming systems activity is focused round a few selected, inter-dependent, inter-related and often inter-linking production systems based on few crops, animals and related subsidiary professions. Integrated farming system involves the utilization of primary produce and secondary produce of one system as basic input of other system, thus making the mutually integrated as one whole unit. There is a need to make effective linkages and complementarities of various components to develop holistic farming system. Integrated farming is defined as biologically integrated system, which integrates natural resources in a regulation mechanisms into farming activities to achieve maximum replacement of off-farm inputs, secures sustainable production of high quality food and other products through ecologically preferred technologies, sustain farm income, eliminates or reduces sources of present environment pollutions generated by agriculture and sustains the multiple function of agriculture. It emphasizes a holistic approach. Such an approach is essential because agriculture has a vital role to play that is much wider than the production of crops, including providing diverse, attractive landscapes and encouraging bio-diversity and conserving wild life. Sustainable development in agriculture must include integrated farming system with efficient soil, water, crop and pest management practices, which are environmentally friendly and cost effective. The future agricultural system should be re-oriented from the single commodity system to food diversification approach for sustaining food production and income. Integrated farming systems, therefore, assume greater importance for sound management of farm resources to enhance farm productivity, which will reduce environment degradation and improve the quality of life of resource poor farmers and to maintain agricultural sustainability. The difference between mixed farming and integrated farming is that enterprises in the integrated farming systems interact eco-biologically, in space and time, are mutually supportive and depend on each other. Hence, integrated farming system is also known as 'Integrated Biosystems'. The aims of the integrated farming system can be achieved by efficient recycling of farm and animal wastes, minimizing the nutrient losses and maximizing the nutrient use efficiency, following efficient cropping systems and crop rotations and complementary combination of farm enterprises. The various enterprises that could be included in the farming system are crops, livestock, poultry, fishery, sericulture, agro-forestry, horticulture, mushroom cultivation, apiary, crop residue recycling etc. Thus it deals with whole farm approach to minimize risk and increase the production and profit with better utilization of wastes and residues. Integrated farming system is based on the concept that "there is no waste", and "waste is only a misplace resource which can become a valuable material for another product. It may be possible to reach the same level of yield with proportionately less input in the integrated farming and the yield would be more sustainable because the waste of one enterprise becomes the output of another, leaving almost no waste to pollute the environment or to degrade the resource base. Integrated farming meets the potentially conflicting challenges at farm level, in a manner that balances food production, profitability, safety, animal welfare, social responsibility and environmental care. Integrated farming seeks to reinforce the positive influences of agricultural production whilst reducing its negative impacts. It is a common sense whole farm management approach that combines the ecological care of a diverse and healthy environment with the economic demands of agriculture to ensure a continuing supply of wholesome, affordable food. It is not prescriptive because it is a dynamic concept: it must have the flexibility to be relevant on any farm, in any country, and it must always be receptive to change and technological advances.

## Intensive Integrated Farming System (IIFS)

IIFS refers to sustainable production on the one hand and livelihood security on the other, wherein all the components of agriculture, horticulture, forestry, livestock, poultry, fishery can be integrated in a complimentary way besides soil conservation measures, vermicompost, mushroom production, apiculture and liquid manure preparation, etc. It involves agricultural intensifications, diversification, value-addition and intensive use of farm resources. To be ecologically sustainable, such intensification is knowledge-intensive rather than capital-intensive and which replace, to the extent possible, market purchased chemical inputs with farm grown biological inputs. IIFS can meet the needs of poor as well as rich farmers and makes the farmer self-sufficient as well as self-reliant. Hence, the integrated farming system fulfills four major aspects of agriculture *viz.* economic viability, easy adaptability, ecological sustainability and social acceptability. The modern agriculture emphasize too more dimensions *viz.*, time and space concept. Time concept relates to increasing crop intensification in situation where there is no constraint for inputs. In rainfed areas where there is no possibility of increasing the intensity of cropping, the other modern concept (space concept) can be applied. In space concept, crops are arranged in tier system combining two or more crops with varying field duration as intercrops by suitably modifying the planting method. Income through arable cropping alone is insufficient for bulk of the marginal farmers. Activities such as dairy, poultry, fish culture, sericulture, bio-gas production, edible mushroom cultivation, agro-forestry and agri-horticulture, etc., assumes critical importance in supplementing their farm income. It should fit well with farm level infrastructure and ensures fuller utilization of bye-products. Intensive integrated farming system is only the answer to the problem of increasing food production for increasing income and for improving the nutrition of small scale farmers with limited resources.

## Why Integrated Farming System?

- **Shrinkage in area under cropping:** Area under cropping is decreasing day by day due to urbanization, industrialization, population, construction of buildings and highways. As a result there is sharp decline in the per capita carrying capacity of the land. The population of India is expected to increase to 137 and 166 crores, while the cultivable land will decline to 141.3 and 131.3 million respectively, in 2030 and 2050 AD.
- **Small and Fragmented holding:** The average holding of a farm in India has been declining and over 80 per cent of operational holdings are below the size of 1.0 hectare.
- **Seasonal nature of income and employment and out-migration:** Cropping activities in rain fed areas areas are restricted to four months in rainy season. Employment opportunities are scarce in other seasons. This leads to large-scale migration of male farmers to cities in search of work. Round the year employment opportunities should be there to check out-migration from rural areas.
- **Deterioration of resource base:** The ultimate goal of sustainable agriculture is to conserve the human population over a longer period. This can be achieved by seeking the optimal use of internal production inputs in a way that provide acceptable levels of sustainable crop productivity and livestock production resulting in economically profitable return.
- **Household requirement:** A country or state is said to achieve complete food and nutritional security if each and every person is able to consume a minimum quantum and quality of various food ingredients *i.e.* adequate and balanced diet on a regular basis. Other requirements include farming system approach is essential for meeting all these diverse needs from limited land holdings of small and marginal farmer.

## Advantages of Integrated Farming System

- **Productivity:** IFS improves space utilization and provides an opportunity to increase economic yield per unit area per unit time by virtue of intensification of crop and allied enterprises. The system improves soil fertility and soil physical structure and using cover crop and organic compost. IFS also reduces weeds, insect pests and diseases from appropriate crop rotation. IFS also ensures rejuvenation of system productivity.
- **Profitability:** Use waste material of one component at the least cost. Thus reduction of cost of production and form the linkage of utilization of waste material, elimination of middleman interference in most input used. Working out net profit B/C ratio is increased.

- **Sustainability:** Organic supplementation through effective utilization of by products of linked component is done thus providing an opportunity to sustain the potentiality of production base for much longer periods. IFS avoids deforestation and thus provides sustainability to the ecosystem.
- **Balanced Food:** Different components of varied nature enables to produce diversified products and different sources of nutrition. IFS ensures food and nutritional security.
- **Environmental Safety:** In IFS waste materials are effectively recycled by linking appropriate components, thus minimize environment pollution and maintain agro-ecological equilibrium.
- **Recycling:** Effective recycling of crop residues, livestock wastes and other unutilized resources.
- **Adoption of New Technology:** Resource rich farmers fully utilize technology. Money flow round the year from different components also gives an inducement to the small/marginal farmers to go for the adoption technologies.
- **Saving Energy:** To identify an alternative source to reduce our dependence on fossil energy source within short time. Through effective recycling technique the organic wastes available in the system can be utilized to generate biogas. Energy crisis can be postponed to the later period.
- **Meeting Fodder Crisis:** Every piece of land area is effectively utilized. Plantation of perennial legume fodder trees on field borders helps in fixing the atmospheric nitrogen. These practices will greatly relieve the problem of non – availability of quality fodder to the animal component linked.
- **Solving Fuel and Timber Crisis:** Linking agro- forestry appropriately the production level of fuel and industrial wood can be enhanced without determining effect on crop. This will also greatly reduce deforestation, preserving our natural ecosystem.
- **Employment Generation:** Combing crop with livestock enterprises would increase the labour requirement significantly and would help in reducing the problems of under employment to a great extent. IFS provide enough scope to employ family labour round the year.
- **Agro-industries:** When one of produce linked in IFS are increased to commercial level there is surplus/value/adoption leading to development of allied agro-industries. Thus IFS enhances opportunities for agriculture based industries.
- **Increasing Input Efficiency:** IFS provide good scope to use inputs in different component greater efficiency and benefit cost ratio. Hence, IFS ensured less reliance to outside inputs – fertilizers, agro-chemicals, feeds, energy etc.
- **Income Rounds the year:** Due to interaction of enterprises with crops, eggs, milk, mushroom, honey, cocoons silkworm. Provides flow of money to the farmer round the year. Thus, IFS improves standard of living of the farm families in spheres of food, clothing, shelter, health and education.

## Scope of Integrated Farming System

- Farming system enterprises include crop, livestock, poultry, fish, agro-forestry etc.
- A combination of one or more enterprises with cropping, when carefully chosen, planned and executed, gives greater dividends than a single enterprise, especially for small and marginal farmers.
- Farm as a unit is to be considered and planned for effective integration of the enterprises to be combined with crop production activity.
- Integration of farm enterprises to be combined on many factors such as:
- Soil and climatic features of the selected area.
- Availability of resources, land, labour and capital.
- Present level of utilization of resources.
- Economics of proposed integrated farming system.
- Managerial skill of the farmer.

## Methodology to Organize Farming Systems under On-Farm Conditions

- Farm selection: Select the agro-ecological zone in which FSR is to be initiated. If necessary, further divide this agro-ecological zone to identify specific farming situation.
- Selection of villages and farmers: Select the village in each farming situation comprising marginal/small and medium/large farmers. Selection of village and farmers should be random so as to represent all farming community of the target area.
- Diagnosis of constraints in increasing farm productivity: Carry out survey through rapid rural appraisal. Prepare an inventory of farm resources and support services. Identify the production constraints.
- Research, design and technology generation and adoption
- Technology transfer and diffusion of improved farming systems within recommended domain.
- Impact of technology of improved farming system – productivity, economic returns, energy input – output, employment, equity (gender issue) and environment.

## Principles of Integration

In agriculture, crop production is the main activity. The income obtained from crops may hardly be sufficient to sustain the farm family throughout the year. Assured regular cash flow is possible when the crop is combined with other enterprises. Judicious combination of enterprises, keeping in view of the environmental conditions of a locality will pay greater dividends. At the same time, it will also promote effective recycling of residues/wastes. Hence, the following principles of integration need to be considered for a successful farming system.

- ☆ The system should produce sufficient high quality food, fiber, fodder and industrial raw material
- ☆ The system should meet the demands of the society
- ☆ The system should maintain a viable farming business
- ☆ The system should care for the environment
- ☆ The system should sustain the natural resources

## Ideal Situations for Introduction of Farming System

- ☆ The farmer wants to improve the soil quality and fertility
- ☆ The farm household is struggling to buy food or below the poverty line
- ☆ The farmer is seeking to maximize profits on existing holding
- ☆ The farmer is looking to reduce chemical control methods
- ☆ The farmers wants to gain profits from allied activities
- ☆ The farm is being eroded by wind or water
- ☆ Chemical inputs are expensive or unavailable
- ☆ Soil fertility has been decreased as a result of inorganic fertilizer use
- ☆ The farm has been suffering from water scarcity
- ☆ The land is fragmented
- ☆ The community wants to reduce pollution
- ☆ The community wants to maintain biodiversity

**Major Components of Integrated Farming Systems**

| *A. Crop* | *B. Livestock and Poultry* | *C. Fishery* | *D. Secondary Agriculture* |
|---|---|---|---|
| Cereals | Cattle | Composite fish culture | Bee keeping |
| Pulses | Buffalo | Fingerling production | Mushroom cultivation |
| Oilseeds | Pig | Paddy cum fish culture | Food processing |
| Fruits | Goat | | Vermicomposting |
| Vegetables | Sheep | | Biogas production |
| Spices | Chicken | | Azolla cultivation |
| Plantation crops | Duck | | Sericulture |
| Flowers | | | Moriculture |
| Fodder/forage crops | | | |
| Agro-forestry | | | |
| Sugarcane | | | |
| Fibre crops | | | |

An IFS can be developed by integrating the above four components (A+B, A+C, B+C, A+D, B+D, C+D, A+B+C, A+B+D, A+C+D, B+C+D, A+B+C+D). Selection of crops, livestock, fishery and secondary agricultural activities depend on farmers' preference, suitable agro-climate, available technology and marketing facility.

## A Conceptual Framework of IIFS Integrating Various Components

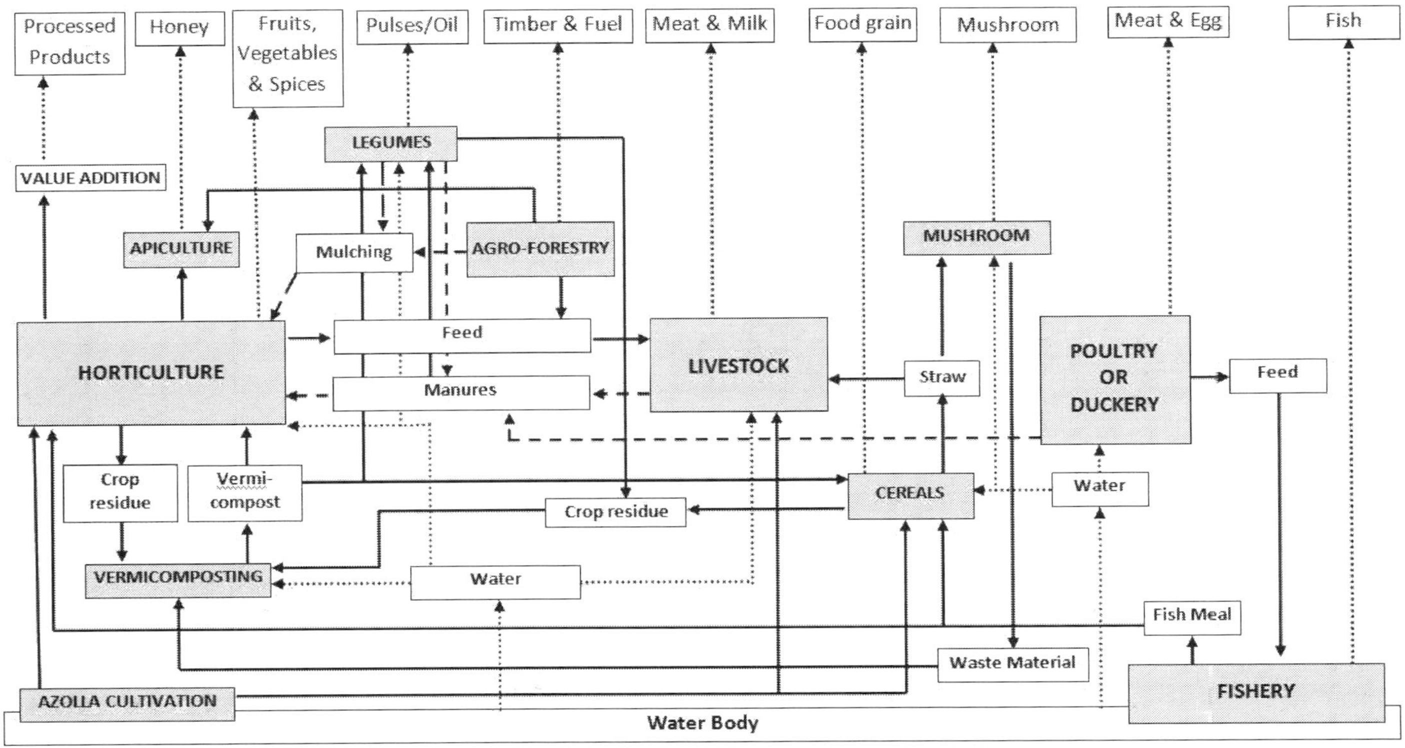

The conservation of natural resources employing the modern concepts of farming systems is essential for sustainable agricultural development and ensuring greater livelihood securities to the poor people of ecologically handicapped areas. Hence, natural resource management (NRM) is an integral part of IFS. NRM refers to the management of natural resources such as land, water, soil, plants and animals, with a particular focus on how management affects the quality of life for both present and future generations. Natural resource management is congruent with the concept of sustainable development, a scientific principle that forms a basis for sustainable global land management and environmental governance to conserve and preserve natural resources. Natural resource management specifically focuses on a scientific and technical understanding of resources and ecology and the life-supporting capacity of those resources. NRM issues are inherently complex as they involve the ecological cycles, hydrological cycles, climate, animals, plants and geography etc. All these are dynamic and inter-related. A change in one of them may have far reaching and/or long term impacts which may even be irreversible. In addition to the natural systems, NRM also has to manage various stakeholders and their interests, policies, politics, geographical boundaries, economic implications and the list goes on. It is very difficult to satisfy all aspects at the same time. This results in conflicting situations. However, following measures can be undertaken for efficient management of natural resources in an integrated farming system.

| *Soil Conservation* | *Water Conservation* |
|---|---|
| ☆ Contour bunding | ☆ Micro-watershed |
| ☆ Bench terracing | ☆ Zingg terracing |
| ☆ Half moon terracing | ☆ Micro-watershed |
| ☆ Bunding (contour, graded) | ☆ Community pond |
| ☆ Gully development / Gully plugging | ☆ Water harvesting (rain, runoff, roof etc.) |
| ☆ Contour trench | ☆ Dew harvesting |
| ☆ Hedge row planting | ☆ Gully plugging |
| ☆ Cover cropping | ☆ Check dam |
| ☆ Intercropping | ☆ Catch pit |
| ☆ Strip cropping | ☆ Scooping |
| ☆ Alley cropping | ☆ Contour trench |
| ☆ Mulching (Horizontal/stubble) | ☆ Percolation tank |
| ☆ Grass waterway/Vegetative barriers | ☆ Mulching (Vertical) |
| ☆ Planting across the slope | ☆ Micro irrigation |
| ☆ Conservation tillage | ☆ Conservation tillage |
| ☆ Manuring/Integrated nutrient management | ☆ Furrow management (ridge and furrow, tied ridging, board bed furrow, dead furrow etc.) |
| ☆ Soil amendments | ☆ Compartmental building |
| ☆ Biological enrichment of soil | ☆ Multiple use of water |

| *Soil Conservation* | *Water Conservation* |
|---|---|
| ☆ Windbreaks, Shelterbelts and soil binding crops | ☆ Perennial horticulture/Agro-forestry |
| ☆ Application of chemicals for aggregate stability (*e.g.* Poly vinyl alcohol, bitumen) | ☆ Organic agriculture |
| ☆ Reclamation of cultivable land | ☆ Revitalization of Traditional Water harvesting structures |

## Integrated Natural Resource Management (INRM)

A process of managing natural resources in a systematic way, which includes multiple aspects of natural resource use (biophysical, socio-political, and economic) meet production goals of producers and other direct users (*e.g.*, food security, profitability, risk aversion) as well as goals of the wider community (*e.g.*, poverty alleviation, welfare of future generations, environmental conservation). It focuses on sustainability and at the same time it tried to incorporate all possible stake holders from the planning level itself, reducing the possible conflicts in future. The conceptual basis of INRM has evolved in recent years through the convergence of research in diverse areas such as sustainable land use, participatory planning, integrated watershed management, and adaptive management. INRM is being used extensively and been successful in community based natural management.

## Community Based Natural Resource Management (CBNRM)

The CBNRM approach combines conservation objectives with the generation of economic benefits for rural communities. The three key assumptions being that; locals are better placed to conserve natural resources, people will conserve a resource only if benefits exceed the costs of conservation and people will conserve a resource that is linked directly to their quality of life. When a local people's quality of life is enhanced, their efforts and commitment to ensure the future well-being of the resource is also enhanced. CBNRM is based particularly on advocacy by nongovernmental organizations working with local groups and communities, on the one hand, and national and transnational organizations, on the other, to build and extend new versions of environmental and social advocacy that link social justice and environmental management agendas with both direct and indirect benefits observed including a share of revenues, employment, diversification of livelihoods and increased pride and identity. CBNRM has raised new challenges, as concepts of community, territory, conservation, and indigenous are worked into politically varied plans and programs in disparate sites.

## Suggested Strategies

- ☆ To ensure timely supply of quality inputs such as seed/planting material and other agri-inputs.
- ☆ To undertake basic and strategic research on production technologies for improving agricultural resource use efficiencies in farming system mode.

- To undertake on-farm testing, verification and refinement of system-based farm production technologies.
- To develop capacity building of stakeholders in Integrated Farming Systems through training and demonstration.
- Optimization of resource inflow and input-output relationships in different IFS models through decision support systems.
- On-farm evaluation and refinement of low-cost storage and value addition tools and techniques for vegetables, fruits, milk and other farm products.
- Advance knowledge and skill development of researchers, extension functionaries and rural entrepreneurs for important areas of activities.
- Location-specific options/interventions for crop-diversification through substitution/intensification/interruption and involving high-value low volume crops and matching production technologies.
- Site-specific nutrient management/balanced nutrient supply systems for intensive cropping systems.
- Development of integrated plant nutrient supply systems for higher productivity and resource sustainability, especially in intensively cultivated areas.
- Development of cost-effective technologies for *in-situ* crop residue management.
- Identification and standardization of new cropping systems appropriate for resource conservation techniques and protected agriculture.
- Dissemination of knowledge on organic farming through Krishi Vigyan Kendra, field demonstrations and mass media (print, electronic and web).
- Assessment of credit delivery system and insurance in agriculture.
- Establishment of farmer-friendly IFS information kiosks in vernacular languages at block level.
- Creation of suitable e-communication platform between farmers and scientists.

Sustainable development on our planet cannot be achieved without a major contribution from agriculture. People must be fed, and agriculture has to face the challenge of producing sufficient food for a rapidly growing world population whilst maintaining the world's fragile resources. Modern farming systems have evolved to meet this need in a way that combines the essential requirements of profitability and productivity. Sustainable development must encompass food production alongside conservation of finite resources and protection of the natural environment so that the needs of people living today can be met without compromising the ability of future generations to meet their own needs. Economic and ecological access to food could be only ensured by adopting farming system approach consisting of change from commodity-based to resource-based planning and integrated use and management of land, water and human resources to maximize income and employment. The primary goals of farming system is to maximize the yield of all component to provide study and stable income at higher level, rejuvenation of systems productivity and achieve agro-ecological equilibrium. Biotech stress management through natural cropping systems management and reducing the use of fertilizers and other harmful agro-chemicals to provide pollution free, healthy produce and environment to the society. Integrated farming system has the advantages of increasing economic yield per unit area per unit time, profitability, sustainability and provides balanced nutritious food for the farmers, pollution free environment and provide opportunity for effective recycling of one product as input to other component, money round the year and solve the energy, fodder, fuel and timber crisis, avoids degradation of forests and enhance the employment generation, increase input use efficiency and finally improve the livelihood of the farming community. Integrated farming systems have emerged as a well-accepted, single window and sound strategy for harmonizing simultaneously joint management of land, water, vegetation, livestock and human resources. A number of such illustrations can be given emphasizing the greater advantage of integrated farming system in generating technologies aimed at combating land degradation. It is this approach that can lead to a quantum jump in the productivity on a sustainable basis and ensure better livelihood securities to the people in fragile ecosystems.

# 2

# Farming System in Manipur

Manipur is often described as "A Little Paradise on Earth". Sometimes it is called as "A Flower on Lofty Height" also. Truly, the state is full of scenic beauty with romantic valley of colourful grasses and flowers against a backdrop of Blue Mountains

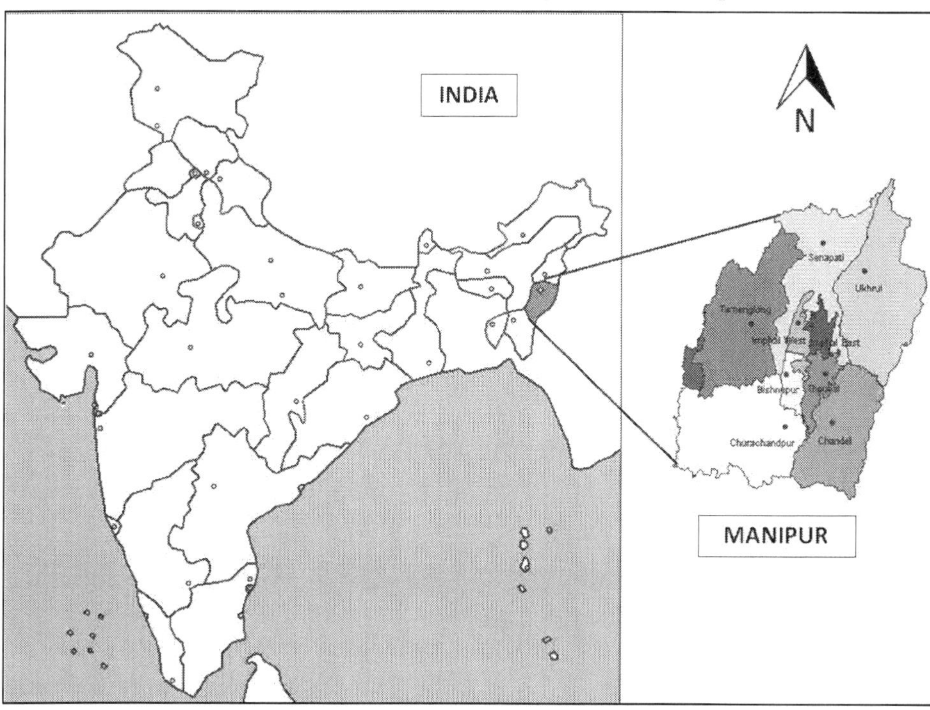

covered with dark forest, just like an adobe of angles under the floating clouds in the blue sky. The state lies between 92°58′E to 94°45′E longitude and 23°50′N to 25°42′N latitude. Altitude varies from 40 to 3114 m above mean sea level. Out of total geographical area of 22,327 sq km, 90 per cent area is under hill tract and rest is central valley. The size of the cultivated area is about 7.41 per cent of the total geographical area of the State. The agriculture in the North Eastern Region of India is characterized by rugged terrain, wide variations in slopes and altitude, community land systems and conventional cultivation practices. The agricultural production system in this region is mostly rainfed, monocropped and at subsistence level. Use of local varieties, negligible use of agro-chemicals, low moisture retention capacity of upland soil and lack of irrigation facilities along with traditional management practices have resulted into low productivity of crops and low cropping intensity in the region. Problems of high rainfall (12.10 per cent of country's total precipitation), soil acidity, aluminium toxicity in upland and iron toxicity in valley land have added to the problem of low agricultural productivity in the region. The soils of the region are usually rich in organic matter and acidic to strongly acidic (pH 4.5-5.0) in reaction. Most of the hill soils are shallow in depth except in valleys and plateaus.

**The Brief Demographic and Economic Scenario of Manipur**

| *Demography* | *Geographical Area (ha)* | *Population* | *Growth* | *Sex Rate (per cent)* | *Literacy Ratio* | *Density (per cent)* | *Land : Man Ratio* |
|---|---|---|---|---|---|---|---|
| 2011 Census | 22,32,700 | 27,21,756 | 18.65 | 987 | 79.85 | 122 | 0.82 |

**Source**: Population Census 2011, Govt. of India.

| *Economy (2012-13)* | *Current Prices* | *Per cent Growth Over Previous Year* | *Constant (2004-05) Prices* | *Per cent Growth Over Previous Year* |
|---|---|---|---|---|
| Gross State Domestic Product (GSDP) | 11713 | 12.52 | 8073 | 7.14 |
| Net State Domestic Product (NSDP) | 10436 | 12.47 | 7248 | 7.17 |
| Per Capita Net State Domestic Product | 36290 | 10.42 | 25205 | 5.23 |

**Source**: NEDFi Databank (http://databank.nedfi.com/content/north-east-resources-databank)

Improved agricultural technologies even when as technically sound are of limited value if they are not accepted by the farmers. New agricultural technologies developed and transferred by the conventional research and extension efforts, have not been adopted uniformly by the farmers operating under widely varying agro-climatic and socio-economic environments when adequate attention is not paid to the agro-climatic and socio-economic milieu in which the farmers operate. Farmers generally seek those technologies which increase their income while keeping risk within reasonable bounds under their own circumstances and management practices. The traditional technology generation and transfer models have been found to be inadequate to meet the specific needs of vast majority farmers. In order to develop and transfer more appropriate technologies to meet the specific client demands, emphasis on farmer's participatory, systemic research and extension effort is now gaining momentum in India and that too in the eastern part of the country. Attempts have been made to integrate the desirable features of farming system research into the mainstream agricultural research so that the technologies developed are relevant, client-oriented and location specific. In North East India crop productivity trends have been below the Indian average for most of the cereal and pulses crops, and far below their potential yield. About 85 per cent of the farmers are small and marginal but sharing only 50 per cent of the land. The 80 per cent of operational farm holdings in India as well as in the eastern India is less than one hectare. Integrated farming system is a reliable way of obtaining high productivity with substantial nutrient economy in combination with maximum compatibility and replenishment of organic matter by way of effective recycling of organic residues/wastes etc. obtained through integration of various land- based enterprises. Farming system approach requires involvement of agriculture, horticulture, soil conservation, forestry, fisheries, animal husbandry (piggery and poultry) and allied enterprises (apiculture etc). Integrated farming system works as a set of systems. From crop production, farmers were getting straw and stover which was used for compost preparation. This compost was inturn used for the production of crops, vegetable and fruits and was also used for fish production. In the same way, poultry and piggery manure from poultry and piggery farming was used for the aforesaid components. The excess produce or waste materials from horticulture are used it as feed materials for piggery. In integrated farming system, rain water harvesting is playing a pivotal role for sustainable production especially in rainfed farming system. This water is used as a life saving for different components of integrated farming system. Thus, IFS is best suited for sustainable production and livelihood development in the NEH Region.

In North Eastern Region, farmers work hard but do not make profit, because there is very little left after they pay for all inputs (seeds, livestock breeds, fertilizers, pesticides, energy, feed, labour, etc.). The emergence of Integrated Farming Systems (IFS) has enabled us to develop a framework for an alternative development model to improve the feasibility of small sized farming operations in relation to larger ones (Ansari *et al.*, 2013, 2014, 2015 and Ravisankar *et al.*, 2006). Integrated farming system is a commonly and broadly used word to explain a more integrated approach to farming as compared to monoculture approaches. It refers to agricultural systems that integrate livestock and crop production or integrate fish and livestock and may sometimes be known as Integrated Bio systems. In this system, an inter-related set of enterprises is used so that the "waste" from one component becomes an input for another part of the system, which reduces cost and improves production and/or income. IFS works as a system of systems (Chan, 2006 and Prakash *et al.*, 2015). IFS ensure that wastes from one form of agriculture become a resource for another form. Since it utilizes wastes as resources, we not only eliminate wastes but we also ensure overall increase in productivity for the whole agricultural systems. We avoid the environmental impacts caused by wastes from intensive activities such as pig farming. The survey has been done by ICAR Research Complex for NEH Region, Manipur

Centre, Imphal about the socio economic as well as agriculture related study of the hilly farmers. On the basis of their problems and prospects, ICAR Research Complex for NEH Region, Manipur Centre, Imphal has been developed various integrated farming system model for distinguish ecosystems of Manipur. It will impart knowledge about the methodology, benefits and component selection at the farmer's field to increase the livelihood of the resource poor farmers with limited land and other resources.

## Predominant Farming System

In Manipur, farmers grow crops and rear livestock, but there is no integration in a way which ensures that the waste product of enterprise is used as input for the other. However, following farming systems have been observed in the state.

### Rice Based Farming System

In the state, rice is the predominant crop grown by the farmers. After the harvest of paddy, there is scarcity of water, due to which the land remains fallow. However, due to efforts of ICAR and its KVKs, the zero-tillage of rapeseed mustard (var. M-27) has become very popular and the farmers are practising this rice-rapeseed based cropping system. The area under zero-tillage of rapeseed mustard is about 15000 hectares which includes all the valley districts and hill district especially Tamenglong and Churachandpur. In some districts like, Thoubal, Bishnupur, Imphal East and some pockets of Imphal West, farmers grow vegetables after rice with assured irrigation facility. The other farming system prevalent in the valley to a limited extent are Rice-Pulse (Pea, Black gram) + Cow, Rice-Rabi Maize + Cow, Rice-Vegetable + Pig, Rice-Vegetable + Poultry, Rice-Potato + Cow, Rice + Fish + Pig and Rice-Rice-Pulse + Cow etc. In some parts of Ukhrul district, farmers are practising Paddy-Legume crop rotation.

### Paddy Cum Fish Culture

Paddy cum fish culture is one of the most important and widely practised farming system in Manipur, especially Bishnupur, Imphal West and Thoubal in valley and Churachandpur, Ukhrul, Tamenglong and Senapati in hills. This is basically rice based cropping system where paddy is grown on terraces. Bunds and terraces control soil erosion, loss of top soil and nutrients. Paddy is basically high water requiring crop, though rain water is sufficiently available but where paddy cum fish culture is taken in this system, water is brought from high hills by diverting through main channels/sub channels ensuring 10-15 cm water depth. Sometimes bamboo channels are also used to divert/carry water to terraces. Paddy is the major crop but on bunds colocasia and yams are also cultivated. Fish crop is traditionally raised only from the paddies of rain fed lowlands (both shallow and deepwater). Fish culture in small water ponds dig out in the middle of terraces is used for rearing of common carps to fetch additional income and water management. This system is basically followed by farmers of all socio economic categories. Adoption pattern for each crop/breed/ other technology under the FS: The adoption pattern of the various components is traditional and primitive. In many areas, irrigation-fed rice fields have also been adapted locally by the farmers to include fish farming. Traditional rice-fish farming systems have an important socioeconomic part in the life of the farmers and fishers in the region. The indigenous rice-fish farming practices prevailing among the farmers can be categorized as (a) rice field capture fishery systems (b) wild aquatic cropping systems (c) mountain valley rice fish farming system and (d) running water terrace rice-fish farming systems. In Oinam village of Bishnupur district and Mayang Imphal area of Imphal West district, farmers are practising Makhana + Fish Culture, which they have found very profitable.

## Horticulture Based Farming System

In valley districts, especially Thoubal, Bishnupur and Imphal East, farmers are practising vegetable based farming system. They grow different vegetables simultaneously. In some areas, Vegetables + Poultry based system has also been observed. Kakching and Wabagai area of Thoubal district are known for 'Vegetable + Vegetable' and 'Vegetables + Spices' based farming system. In hilly areas, farmers grow different types of vegetables in small areas for their household consumption. They also grow banana and colocasia and use the banana pseudostem and colocasia leaves and stem for feeding pigs. In Chandel district, sweet potato is grown for both household consumption and pig feed. In some parts of Senapati district, farmers plant turmeric under the shade of forest trees. In Ukhrul and Tamenglong, king chilli are grown under the shade of bamboo plants. In Churachandpur district, farmers grow ginger and chilli in their passion fruit plantation. Farmers of the hill districts also grow colocasia and maize as intercrop in their orange and lemon orchards, which is a very harmful practice. In some parts of Tamenglong district, Papaya + Vegetables and Banana + King Chilli intercropping system has also been observed.

## Livestock Based Farming System

In the valley, especially in the urban and peri-urban areas, majority of the farmers do not rear livestock. However, the farmers in rural areas rear poultry, ducks, local cow, pigs etc. In the periphery of Loktak lake in Bishnupur district, some of the farmers have integrated duckery, fishery with deep water rice. In Thoubal and Bishnupur district, rearing of buffalo is common. Mithuns are being maintained in semi-wild condition mainly in Senapati and some parts of Chandel and eastern part Ukhrul districts. In all the hill districts piggery is very popular. In Imphal West and Imphal East districts, some tribal farmers practice piggery + fish culture. Both, in the valley and hill regions, farmers have taken up backyard poultry rearing to meet their household needs and supplement their income. The fodder supply for the livestock in the valley is met from locally available material, kitchen waste etc.

## Agri-Silvi-Horti-Pastural Farming System

Agri-silvi-horti-pasture is an indigenous system of farming practiced by the local people from time immemorial. In this system, crops and livestock component are taken along with natural forest cover. Maize is major cereal crop grown in the system, however Banana, Mandarin, Passion fruit, Pear, Peach, Plum are the main horticulturalcrops. In certain pockets cabbage and beans are also being taken as mixed crop with maize. Forest and Fodder trees like *Ficus* spp, *Bauhinia* spp., *Legistroma* etc. Fodder grasses like Broom grass, Napier and Guinea grass are also grown with

Livestock component Cattle. Mono cropping and mixed cropping system is followed in Agrisilvihortipastural system. The land use pattern of the farming system is divided into Forest area which normally being used for grazing of cattle. This system is practiced on village community land.

## Shifting Cultivation

Shifting cultivation, locally known as *Jhuming*, is a predominant system of farming in hill districts of Manipur. With the rising population, the cycle of shifting cultivation which was 15-20 years earlier has reduced to less than 5 years, resulting in degradation of forests in these areas. The area of shifting cultivation (particularly jhum paddy) is more than 52.44 per cent of the gross cropped area of the hill districts, 23.23 per cent of the total gross cropped area and 3.53 per cent of the total geographical area of the state. An additional area of 20.13 thousand hectares has been reported as abandoned jhum land (Wasteland Atlas of India, 2011). Farmers cut and burn the fallow vegetation from November onwards up to January. Thereafter, they grow rice and vegetables in their jhum land. Different types of mixed cropping have been observed in the jhum areas like Upland paddy + Vegetables (Cabbage) in Chandel, Upland paddy + Spices (King Chilli) in Tamenglong etc. The improved method of shifting cultivation is developed by the Tribals of Machi Block of Chandel district in Manipur, because of the acute problem for acquiring land for shifting cultivation. This system is solely dependent upon rainfall. The adoption pattern of the various components is traditional and primitive. In fact, shifting cultivation is not in any way a farming system, rather it can be considered as a farming practice to meet the household needs of the farm family. The choice of crops differs widely from district to district or even within the district. Sometimes, mixed cropping of 7-8 different types of crops have also been observed in a small jhum land. Maximum area under jhum is contributed by Churachandpur, followed by Tamenglong. In Ukhrul district, shifting cultivation is mostly prevalent in southern part. As the socio-economic condition of the hill farmers are very poor, majority of them do not have enough capital for land development (vegetation clearing, terracing etc.) to start settled cultivation. In some parts of Ukhrul, farmers have shifted from wetland cultivation of paddy in terraces to jhum land, because the productivity of paddy in the terraced land has decreased. This is probably due to continuous cultivation of paddy without any application of manures and fertilizers which has led to exhaustion of nutrients in the soil. The jhum land is left to rejuvenate on its own for a few years after the farmer abandons the cultivation on it. Till date organized initiative has not been undertaken to rejuvenate the abandoned jhum lands. Hence, shifting cultivation still remains the way of life for the hill farmers of Manipur.

## Agroforestry

A survey on the agroforestry system in five hill district and four valley districts of Manipur revealed the problems of degradation of natural resources, mainly forest soil and water. Jhumming in the hills, population pressure, land use pattern, ignorance, lack of proper transport and communication facilities contributed towards the low agricultural production status. In the valley, only areas of religious importance and some pockets are conserved with forest tree. Inspite of this some need based systems are identified. Multipurpose tree species *viz., Parkia roxburghii, Ailanthus excels, Alnas nepalenesis, Quercus* in combination with various agricultural and horticultural crops are found promising in developing sustainable agroforestry system for this region. About 28 tree species are identified as suitable components for agroforestry system in Manipur. A combination of *Alnus nepalensis* + large cardamom, *Ailanthus excels/Albizia lebeck + coffee, Parkia roxburghii,/Cinnamomum* sp. + turmeric/ginger, *Emblica officinalis* + *Quercus* sp etc are found promising.

Farming system approach to agricultural research and development efforts would accelerate agricultural growth of Manipur and thereby providing leverage for transforming poverty prone rural areas to a prosperous state by strengthening rural economy.

# 3

# Resource Inventory

The importance of agriculture for the Indian society can hardly be over-emphasized, as its role in economy, employment, food security, national self reliance and general well being, does not need reiteration. The need for timely and reliable information on crop area and production for tactical and strategic decision making by all stakeholders in agriculture, such as producers, processors, resource managers, marketing, finance and the government is well known. With the global shift in the market economies, reliable agricultural information has gained more importance than ever before. Use of this information could be made for a number of applications such as crop inventory, crop production forecasts, drought and flood damage assessment, range and irrigated land monitoring and management. Crop inventory as well intervention of appropriate technology is a major issue for increasing livelihood of rural farmers in Manipur. The word resource has originated from French dialect *resource*, 'rise again, or recover' and inventory has originated from Latin word *inventarium* 'a list of what is found'. So resource is any physical or virtual entities of limited availability that help one earn a living. Resources are very useful to produce goods so they possess economic value. Inventory is a list for articles, goods and materials available in the stock at particular period of time. Thus resource inventory can be defined as "list of contents of a physical or virtual entity required for production of goods having economic value". Resources have three main characteristics; quantity, utility and availability.

Manipur is bounded by Myanmar (Burma) on the east; Nagaland on the north; Assam on the west; and Mizoram on the south. It covers an area of 22,327 sq. kms. Physiographically Manipur is one of the most enchanting places in the country. It can be broadly classified into three well defined geographical region *viz.*, the hilly region, the central plain or the Manipur valley and the Jiribam plain. The hilly terrains of Manipur stretches mainly from North to South or North-South parallel folds. The Manipur valley is situated in the central part in Manipur covering an area of 1843 $m^2$ which account for about 10 per cent of total land area of the state having an oval shaped elevated plain surrounded by hilly on all sides and extended about 58 km from North to South and 31 km from East to West. About 57.20 per cent of the total population (28,55,794 persons) of the state resided in the central plain of Manipur. The whole valley slopes from North to South. The Jiribam plain lies beyond the Manipur Western hill tract covering an area of 1 per cent of the land area. The literacy rate of the state is 79.21 per cent. Agriculture is the back bone of Manipuri's economy. About 80 per cent of the state total population is engaged in agriculture and allied activities. There are vast tracts of fertile land in Manipur with a gross cropped area of 3,39,290 ha which account for about 15.19 per cent of the total land areas of the state and the about 65.67 per cent of the gross cropped areas is utilize for paddy cultivation. The tribal people living in the hills usually practice Jhum cultivation. The mean cropping intensity of the state is 145.16 per cent only. Rice is the stable fruit crop in Manipur which is grown on a large scale. Paddy is the largest cultivated crop both in hills and in plans. The other important food crops of the state include wheat, maize, oilseeds, pulses and potato. Potato, ginger, gram, cauliflower, cabbage, mustard, turmeric, and other spices crop are the major crops of Manipur. Among the horticultural crops, pineapple, banana, orange, lemon passion fruit and papaya are the major fruit crops. Almost all kinds of vegetables especially cucurbits, legumes and cole crops are grown here along with indigenous vegetable crops. Turmeric, ginger and chilli are the major spices. The whole state is a treasure trove of many indigenous and underutilized horticultural crops including orchids and medicinal plants. The state is bestowed with suitable agro-climate for practicing sericulture. Bamboo, oak, teak, pine, leihao and uningthou etc. cover the forest areas. Manipur has got plenty of fishery resources in the form of lake, reservoir, tanks, ponds, marshy areas, natural basin, water lock areas and canal etc, occupying 56461.00 ha. Cattle, buffalo, pig, goat, mithun and ship are the major livestock. The state has also shown considerable development in the poultry sector through introduction of improved strains.

## Soil and Climate

### Soils

The soil cover can be divided into two broad types, *viz.* the red ferrogenous soil in the hill area and the alluvium in the valley. The soil generally contains small rock fragments, sand and sandy clay and are of varieties. The top soil on the steep slopes is very thin. In the plain areas, especially flood plains and deltas, the soil is of considerable thickness. Soil on the steep hill slopes is subjected to high erosion resulting into formation of sheets and gullies and barren rock slopes. There are four major soil order found in Manipur namely entisol, inceptisol, alfisol and ultisol. In general, soil fertility status is low to medium. The normal pH value ranges from 5.4 to 6.8. Soil group classifications present in the district:

1. Fine Loamy

2. Clay loam
3. Laterite soil
4. Clayey red loamy soil
5. Yellow soil

## Climate

Manipur falls under the monsoon belt of India. The climate of the state is salubrious with annual rainfall varying from 933 mm to 2593 mm. Period from June to September is the monsoon season. Due to varied level of altitude and slopes the climatic of Manipur varies from tropical to subtropical to semi-temperate condition (prevailing at higher altitudes). The winter season (November to February) can be characterized by low temperature, heavy dew fall and occurrence of frost. The temperature varies from 0°C in winter to 36°C in summer months. The state can be divided in two distinct agro-climatic zones:

1. **Central valley and adjoining areas** – The land is flat, soil is heavy to very heavy in texture. In some pockets, light textured soil can also be seen. Elevation ranges from 750 to 950 m above MSL. This region has two broad sub-regions – River banks and surrounding areas and valleys.
2. **Undulating hill terrain** – Soil in this region is light textured, elevation ranges from 1000 to 3000 m above MSL. This region can be divided in to two sub-regions – mid hills (elevation 1000-1500 m from MSL) and high hills (elevation above 1500 m from MSL)

## Status of Agriculture

Agriculture being the main occupation of the people of Manipur, it has an important place in the economy of the state. Agriculture sector contributes a major share to the total state domestic product and provides employment to about 50 per cent of the total workers in Manipur. The size of the cultivated area is about 10.46 per cent only of the total geographical area of the State. Of this total cultivated area, 52 per cent is confined to the valley. In the plains/valleys 'settled or permanent' cultivation is practiced while in the hills 'shifting cultivation is prevalent. The productivity of paddy in valley is higher than at the national level. The hills produced only one fourth of the state's total production. Of the total area under paddy about 15 per cent was under Jhum cultivation during 1991-92, which increased to around 26 per cent in 2001-02 and 35 per cent in 2013-14.

**Land Use Classification of Manipur (2013-14)**

| | |
|---|---|
| A. Total geographical area of the state: | 2,232,700 hectares |
| B. Gross cropped area of the state: | 3,39,290 hectares (15.19 per cent of A) |
| C. Net cropped area of the state: | 2,33,740 hectares (10.46 per cent of A) |
| D. Gross cropped area of hill districts: | 1,50,280 hectares |
| E. Gross cropped area of valley districts: | 1,89,010 hectares |
| F. Net sown area of hill districts: | 1,13,020 hectares |

| | |
|---|---|
| G. Net sown area of valley districts: | 1,20,720 hectares |
| H. Per cent gross cropped area in the hills to the total gross cropped area of the state: | 44.29 per cent |
| I. Per cent gross cropped area in the valley to the total gross cropped area of the state: | 55.70 per cent |
| J. Per cent gross cropped area in the hills to the total geographical area of the state: | 6.73 per cent |
| K. Per cent gross cropped area in the valley to the total geographical area of the state: | 8.46 per cent |
| L. Per cent of net sown area in the hills to the total net cropped area of the state: | 48.35 per cent |
| M. Per cent of net sown area in the valley to the total net cropped area of the state: | 51.64 per cent |
| N. Per cent of net sown area in the hills to total geographical area of the state: | 5.06 per cent |
| O. Per cent of net sown area in the valley to total geographical area of the state: | 5.40 per cent |
| P. Area under horticultural crops: | 93,285 hectares (4.17 per cent of A) |
| Q. Area under forest cover: | 16,99,000 hectares (76.09 per cent of A) |
| R. Total water area (2009-10): | 56,461 hectares (2.52 per cent of A) |
| S. Mean cropping intensity of the state: | 145.16 per cent |
| T. Mean cropping intensity of the hill districts: | 132.97 per cent |
| U. Mean cropping intensity of the valley districts: | 156.57 per cent |

**Source**: (a) Department of Agriculture, Govt. of Manipur; (b) Economic Survey Manipur 2013-14, Department of Economics and Statistics, Govt. of Manipur and (c) India State of Forest Report 2013, Forest Survey of India.

## Major Farming Enterprises in Manipur

### A. Rice

Rice is the most important crop and the staple food of the people in Manipur. The local people prefer to eat glutinous rice (also known as sticky rice) as compared to non-glutinous one. The agro-climatic condition is most suitable for the cultivation of rice in the state. The valley area has an average altitude of 800 meters above MSL and the climate is subtropical and warm in the summer season. This part is the "rice bowl" of the state. In the hilly terrains, rice is grown in the jhum land or in terraces. Paddy is cultivated by traditional or japanese method of planting. Now days, system of rice intensification (SRI) is also becoming popular in some pockets. The state is rich in rice diversity and a good number of local varieties are available in both hill and plain area. Farmers usually grow local varieties; but presently high yielding varieties have gained popularity. Hybrid rice is cultivated by some progressive farmers. Hence, rice contributes a major share of household income in Manipur.

**Area, Production and Productivity of Rice in Manipur (2013-14)**

| Crops | Area ('000 ha) | Production ('000 MT) | Productivity (t/ha) |
|---|---|---|---|
| Pre Kharif Paddy | 27.80 | 91.78 | 3.30 |
| HYV Paddy | 106.18 | 393.25 | 3.70 |
| Jhum Paddy | 78.82 | 82.43 | 1.05 |
| Terrace Paddy | 10.00 | 30.21 | 3.02 |
| **Total Paddy** | **222.80** | **597.67** | **2.68** |

**Source**: Department of Agriculture, Govt. of Manipur.

Of the total geographical area of the state paddy occupies almost 10 per cent area, which is 65.67 per cent and 95.32 per cent of the gross and net cropped area of the state, respectively. Though hill districts occupies almost 40 per cent of the paddy growing area in the state, it merely contributes 18.85 per cent of the total paddy production; whereas, valley area contributes 81.15 per cent of the total production with 60.13 per cent share in total area under paddy. The average productivity of paddy in the valley districts is quite high *i.e.* 3.53 t/ha as compared to the hill districts (1.33 t/ha), which is relatively low. The mean productivity of paddy in Manipur was 2.35 t/ha during 2013-14. The major constrain of rice cultivation are poor nursery raising technique, high seed rate, low yield of local varieties, non availability of seed of high yielding/hybrid varieties in sufficient quantity, lack of suitable varieties for jhum cultivation, limited use or no use of nutrients to the crop especially in hills, improper weed management, insect pest and diseases infestation, no use of organic/ inorganic amendments etc. All the constrains can be overcome with proper and appropriate location specific technologies.

## B. Maize

Maize is the second most important crop grown in Manipur after rice. It is grown in both in hills and in plains. Maize is mainly grown in hill areas with an altitude ranging from 850 to 1800 above msl. Like rice, farmers mainly grow local varieties as the state is also rich in maize diversity. However, farmers started preferring composite and hybrid maize varieties from last couple of years.

**Area, Production and Productivity of the Maize in Manipur (2013-14)**

| Crops | Area ('000 ha) | Production ('000 MT) | Productivity (t/ha) |
|---|---|---|---|
| Kharif Maize | 19.59 | 42.75 | 2.18 |
| Rabi Maize | 6.50 | 15.86 | 2.44 |
| **Total Maize** | **26.09** | **58.61** | **2.25** |

**Source**: Department of Agriculture, Govt. of Manipur

In case of maize, hill districts share the maximum area (76.58 per cent of the total are under maize) and also the production (75.82 per cent of the total maize production) in the state. However, the productivity of maize in hill districts (2.22 t/ ha) is little lower than valley area (2.33 t/ha). The main causes of low productivity of maize in Manipur are unsuitable methods of planting, low yield of local varieties, non availability of seed of high yielding/hybrid varieties in sufficient quantity, high seed rate, cob borer infestation, stem borer infestation and nutrient deficiency and improper nutrition to the crop.

## C. Pulses and Oilseeds

Cultivation and production of pulses and oilseed in Manipur is very less. The major pulse crops grown in Manipur are field pea, rice bean, broad bean, cow pea, mung bean, black gram, red gram etc. Cultivation of lentil has been introduced recently in Thoubal district. Among the oilseed crops groundnut, rapeseed mustard and soybean are most important. Zero tillage cultivation of rapeseed mustard is quite popular in the state.

**Area, Production and Productivity of Pulses and Oilseeds in Manipur (2013-14)**

| *Crops* | *Area ('000 ha)* | *Production ('000 MT)* | *Productivity (t/ha)* |
|---|---|---|---|
| Kharif Pulses | 4.88 | 5.21 | 1.07 |
| Rabi Pulses | 25.50 | 23.44 | 0.92 |
| Total Pulses | 30.38 | 28.65 | 0.94 |
| Kharif Oilseed | 8.42 | 7.16 | 0.85 |
| Rabi Oilseed | 28.50 | 23.85 | 0.84 |
| **Total Oilseed** | **36.92** | **31.01** | **0.84** |

**Source**: Department of Agriculture, Govt. of Manipur

Both valley and hill districts contribute almost equal shares in terms of area and production of both pulses and oilseeds. However, the productivity of pulses and oilseeds is relatively low (0.94 t/ha and 0.84 t/ha, respectively) in Manipur. The main constrains for pulses and oil seed productions are cultivation in highly eroded areas, water scarcity during growing season (as most of the crops grown under rainfed condition), lack of seed of high yielding varieties in sufficient quantity, improper crop management practices, heavy infestation of weeds, insect pests and diseases, farmer's preference to grow grain crops, faulty post harvest management and lack of processing and storage.

## D. Other Field Crops

Among the other field crops, potato is most popular. Wheat and sugarcane is grown in isolated pockets of the state.

**Area, Production and Productivity of Other Field Crops in Manipur (2013-14)**

| *Crops* | *Area ('000 ha)* | *Production ('000 MT)* | *Productivity (t/ha)* |
|---|---|---|---|
| Wheat | 2.25 | 5.62 | 2.50 |
| Sugarcane | 5.85 | 339.31 | 58.00 |
| Potato | 15.00 | 127.00 | 8.47 |
| **Total** | **23.10** | **471.93** | **68.97** |

**Source**: Department of Agriculture, Govt. of Manipur

Non-availability of quality seed/planting material, lack of knowledge on improved package of practices, and unfavourable climatic condition (especially for wheat) are the major bottlenecks.

**Share of Different Districts in Area and Production of Field Crops during 2013-14 in Manipur**

| | | |
|---|---|---|
| **Paddy** | A | Imphal West > Imphal East > Thoubal > Churachandpur > Bishnupur > Tamenglong > Senapati > Ukhrul > Chandel |
| | P | Imphal West > Imphal East > Thoubal > Bishnupur > Churachandpur > Tamenglong > Senapati > Ukhrul > Chandel |
| **Maize** | A | Churachandpur > Senapati > Chandel > Tamenglong > Ukhrul > Thoubal > Bishnupur > Imphal West > Imphal East |
| | P | Churachandpur > Senapati > Chandel > Tamenglong > Ukhrul > Thoubal > Bishnupur > Imphal West > Imphal East |
| **Pulses** | A | Bishnupur > Thoubal > Imphal West > Imphal East > Ukhrul > Tamenglong > Chandel > Churachandpur > Senapati |
| | P | Bishnupur > Thoubal > Imphal East > Imphal West > Ukhrul > Tamenglong > Chandel > Churachandpur > Senapati |
| **Oilseeds** | A | Bishnupur > Thoubal > Ukhrul > Imphal East > Imphal West > Senapati > Tamenglong > Churachandpur > Chandel |
| | P | Bishnupur > Thoubal > Imphal East > Ukhrul > Imphal West > Tamenglong > Senapati > Churachandpur > Chandel |
| **Wheat** | A | Senapati > Churachandpur > Thoubal > Bishnupur > Imphal West > Imphal East |
| | P | Senapati > Churachandpur > Thoubal > Bishnupur > Imphal West > Imphal East |
| **Sugarcane** | A | Bishnupur > Thoubal > Imphal West > Imphal East > Chandel > Churachandpur > Senapati > Tamenglong > Ukhrul |
| | P | Bishnupur > Thoubal > Imphal West > Imphal East > Chandel > Churachandpur > Senapati > Tamenglong > Ukhrul |
| **Potato** | A | Bishnupur > Thoubal > Tamenglong > Churachandpur > Chandel > Imphal West > Senapati > Imphal East > Ukhrul |
| | P | Bishnupur > Thoubal > Tamenglong > Chandel > Churachandpur > Imphal West > Senapati > Imphal East > Ukhrul |

A = Area, P = Productivity.

## E. Horticultural Crops

Horticulture plays an important role in the subsistence farming which is being practised by majority of the small and marginal farmers, especially in hill districts of Manipur. The state is bestowed with all kinds of pre-requisite for successful cultivation of various horticultural crops. The major fruit crops of the state are pineapple, banana, passion fruit, khasi mandarin and kachai lemon. Almost all kinds of vegetables are grown in the state; however, cole crops, legumes and cucurbitaceous crops are mostly preferred. The major spices are turmeric, ginger and chilli. The state is known for king chilli, one of the hottest chilli in the world. There is enormous potential for floriculture sector in the state. Moreover, Manipur is a treasure trove of many underutilized and indigenous horticultural crops, some of them are really rich in bioactive properties and are being used in traditional and ethnic medicines. But, the horticulture sector has still remained unexploited from commercial point of view.

**Area, Production and Productivity of the Major Horticultural Crops in Manipur (2013-14)**

| Crops | | Area (Ha) | Production (MT) | Productivity (MT/Ha) |
|---|---|---|---|---|
| **FRUITS** | Pineapple | 13700 | 136315 | 9.95 |
| | Banana | 6700 | 90450 | 13.50 |
| | Lime/Lemon | 5850 | 52650 | 9.00 |
| | Orange | 5150 | 41200 | 8.00 |
| | Passion Fruit | 9152 | 100580.5 | 10.99 |
| | Others | 13500 | 94500 | 7.00 |
| | **Sub-Total** | **54052** | **515695** | **9.54** |
| **VEGETABLES** | Cauliflower | 2750 | 30250 | 11.00 |
| | Cabbage | 7104 | 84538 | 11.90 |
| | Tomato | 2830 | 28725 | 10.15 |
| | Pea | 5809 | 59426 | 10.23 |
| | Others | 6000 | 60900 | 10.15 |
| | **Sub-Total** | **24493** | **263838** | **10.77** |
| **SPICES** | Chilly | 8950 | 70616 | 7.89 |
| | Ginger | 2850 | 34200 | 12.00 |
| | Turmeric | 1910 | 22920 | 12.00 |
| | Others | 1030 | 5923 | 5.75 |
| | **Sub-Total** | **14740** | **133658** | **9.07** |
| | **GRAND TOTAL** | **93285** | **913191** | **9.79** |

**Source**: Department of Horticulture and Soil Conservation, Govt. of Manipur.

Among the horticultural crops, fruits alone occupied almost 58 per cent of the total area and 56.47 per cent of the total production of in the state; whereas, vegetables occupied 26.25 per cent and 28.89 per cent of the total area and production, respectively. The share of spice crops was 15.80 per cent of the total area and 14.64 per cent of the total production. Hill districts contributes in 57.56 per cent of the total area 57.06 per cent of the total production of horticultural crops in the state; whereas, valley districts contributed 42.44 per cent and 42.94 per cent of the total area and production, respectively. In fruit sector, the maximum area and production is contributed by the hill districts (70.61 per cent and 70.66 per cent of the total area and production, respectively); whereas, valley districts dominates in vegetable sector with 67.23 per cent and 67.27 per cent share of the total area and production, respectively. However, in spice sector the contribution of hill and valley districts is almost equal in case of area (Hill: 50.90 per cent and Valley: 49.10 per cent) and also in production (Hill: 52.60 per cent and Valley: 47.40 per cent). Though the area and production of horticultural crops in the state has been increased significantly over past few years; however, the productivity has still remained low. The full potential of the horticultural sector (especially flowers, ornamental, medicinal and aromatic crops) in Manipur is yet to be realized.

**Share of Different Districts in Area and Production of Horticultural Crops during 2013-14 in Manipur**

| | | |
|---|---|---|
| **Fruits** | A | Tamenglong > Ukhrul > Senapati > Chandel > Thoubal > Churachandpur > Bishnupur > Imphal West > Imphal East |
| | P | Ukhrul > Tamenglong > Senapati > Chandel > Thoubal > Churachandpur > Bishnupur > Imphal West > Imphal East |
| **Vegetables** | A | Thoubal > Imphal East > Bishnupur > Imphal West > Senapati > Churachandpur > Ukhrul > Chandel > Tamenglong |
| | P | Thoubal > Imphal East > Bishnupur > Imphal West > Senapati > Churachandpur > Ukhrul > Chandel > Tamenglong |
| **Spices** | A | Churachandpur > Imphal East > Thoubal > Imphal West > Senapati > Bishnupur > Tamenglong > Chandel > Ukhrul |
| | P | Churachandpur > Imphal East > Thoubal > Imphal West > Senapati > Bishnupur > Tamenglong > Chandel > Ukhrul |

A = Area, P = Productivity

**Crop Calendar of Manipur (Valley)**

| *Crops* | *Jan* | *Feb* | *Mar* | *Apr* | *May* | *Jun* | *Jul* | *Aug* | *Sep* | *Oct* | *Nov* | *Dec* |
|---|---|---|---|---|---|---|---|---|---|---|---|---|
| VALLEY | | | | | | | | | | | | |
| Prekharif Rice | | + | + | + | + | + | + | | | | | |
| Kharif Rice | | | | | | + | + | + | + | + | + | |
| Kharif Maize | | | | + | + | + | + | | | | | |
| Rabi Maize | + | + | + | + | + | | | | | | + | + |
| Kharif Pulses | | | | | | | + | + | + | | | |
| Rabi Pulses | + | + | + | + | | | | | | | + | + |
| Kharif Oilseeds | | | | | | + | + | + | + | + | | |
| Rabi Oilseeds | + | + | | | | | | | | | + | + |
| Sugarcane | | + | + | + | + | + | + | + | + | + | + | + |
| Potato | + | + | + | | | | | | | | + | + |
| Vegetables | + | + | + | + | + | + | + | + | + | + | + | + |
| Spices | + | + | + | + | + | + | + | + | + | + | + | + |

**Crop Calendar of Manipur (Hills)**

| Crops | Jan | Feb | Mar | Apr | May | Jun | Jul | Aug | Sep | Oct | Nov | Dec |
|---|---|---|---|---|---|---|---|---|---|---|---|---|
| HILLS | | | | | | | | | | | | |
| Prekharif Rice | | | + | + | + | + | + | | | | | |
| Kharif Rice | | | | | | + | + | + | + | + | + | |
| Kharif Maize | | | + | + | + | + | + | | | | | |
| Kharif Pulses | | | + | + | + | + | | | | | | |
| Rabi Pulses | + | | | | | | | | + | + | + | + |
| Rabi Oilseeds | | | | | | | | | + | + | + | + |
| Sugarcane | | + | + | + | + | + | + | + | + | + | + | |
| Potato | | + | + | + | + | | | | | | | |
| Vegetables | | | + | + | + | + | + | + | + | + | + | |
| Spices | | + | + | + | + | + | + | + | + | + | | |

## F. Livestock and Poultry

Amongst the allied enterprises of agriculture, livestock and poultry is the important sector in Manipur. Development of animal husbandry is an essential feature as livestock plays a pivotal role in the state's rural economy. A large number of small and marginal farmers, agricultural labourers and other economically weaker section depend up on livestock for gainful employment. the main categories of livestock reared in Manipur are cattle, buffalo, sheep, goat, mithun, pig etc. Rearing of pigs and poultry are important sources of income generating activities, especially in hill districts. Cattle population constitutes 37.62 per cent of the total livestock population. During 2013-14, the state has produced 80.03 thousand tonnes milk, 1162.19 lakh numbers eggs and 25.02 thousand tonnes meat.

**Livestock and Poultry Population in Manipur**

| Livestock and Poultry | Census 2012 | | Livestock and Poultry | Census 2012 | |
|---|---|---|---|---|---|
| | No. (in '000) | Per cent to Total Livestock | | No. (in '000) | Per cent to Total Livestock |
| Cattle | 263.84 | 37.92 | Mules | 0.37 | 0.05 |
| Buffalo | 66.37 | 9.54 | Donkey | 0.13 | 0.02 |
| Sheep | 11.46 | 1.65 | Dog | 161.82 | 98.75 |
| Goats | 65.16 | 9.37 | Rabbit | 2.04 | 1.25 |
| Pig | 277.22 | 39.84 | Fowl | 1928.06 | 77.14 |
| Mithun | 10.13 | 1.46 | Duck | 551.68 | 22.07 |
| Horses and Pony | 1.10 | 0.16 | Turkey and other poultry | 19.78 | 0.79 |

**Source**: 19th Livestock Census - 2012 All India Report, Govt. of India.

## G. Fishery

Fish is the main food item of the majority of the people in the state. The state has no marine fisheries. It has vast potential of fisheries resources comprising ponds, tanks, natural lakes, marshy areas, rivers, reservoirs, submerged crop land, low lying paddy fields etc. The largest source of fish is Loktak lake (230 sq. km. surface area) in Bishnupur district.

**Fishery Resources In Manipur (2009-10)**

| *Particulars* | *Water Area (ha)* |
|---|---|
| Lakes, Reservoirs, tanks, Canals | 13221.45 |
| Water logged marshy and swampy Lands, Beels | 11536.23 |
| Biomass | 8596.50 |
| Submerged crop land | 3480.50 |
| Rivers and Streams | 13888.27 |
| Water logged areas converted in to agricultural lands | 1738.10 |
| Low lying paddy field | 4000.00 |
| **Total** | **56461.05** |

The total water area in Manipur state has shrunk from around 1 lakh hectares in 1990 to around 56.46 thousand hectares in 2009-10. The state has produced 28.54 thousand tonnes of fish and more than 125 million fish seed in 2013-14.

## Climate Change: Trends and Projections

Climate change is a reality and Manipur state has also been affected due to changing climatic scenario. Trend analysis of weather variables of Imphal region under National Innovations on Climate Resilient Agriculture (NICRA) revealed that the mean annual maximum temperature (1954-2014) has shown increasing trend (0.1°C per decade). The mean annual minimum temperature has also increased significantly (3°C per decade). The trend of annual temperature and rainfall in Imphal region are presented in the following figures (ICAR, Manipur Centre, 2015).

The total annual rainfall recorded during the period from 1954 to 2014 showed increasing trend (23.5 mm per decade); however, decreasing trend has been observed in February, June, July and November (-10.8 to -0.1 mm per decade). The mean annual rainy days (1954-2013) has shown increasing fashion. Decreasing trend in total monthly rainy days has been observed in January, February, April, June and July. The mean annual maximum relative humidity (RH) increased significantly (3.8 per cent per decade) during the period from 1985-2013; whereas, the mean annual minimum RH (1985-2013) shown a decreasing trend (-1.6 per cent per decade). The mean monthly maximum RH has shown significantly increasing trend (3.0 to 5.6 per cent per decade) throughout the year. Similarly, decreasing trend has been observed in mean monthly minimum RH (-3.2 to -0.6 per cent per decade) in all months except May (1 per cent per decade).

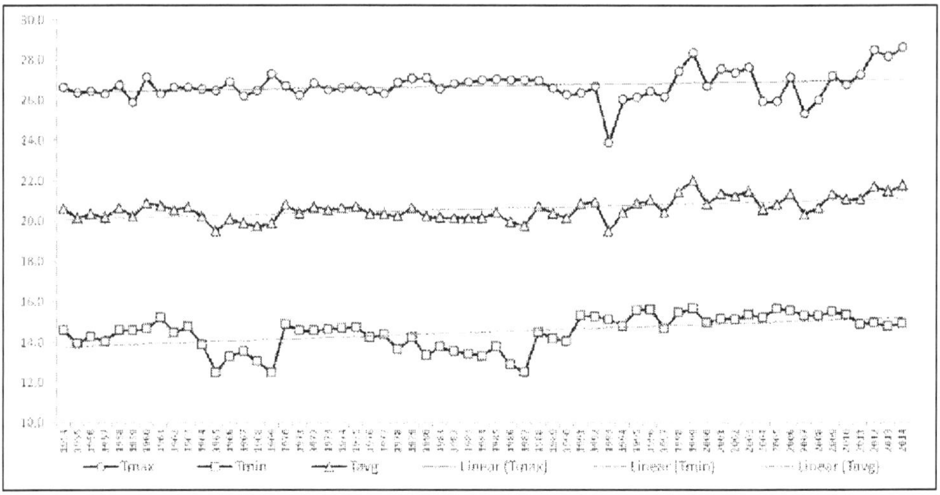


**Annual Temperature of Imphal Region, Manipur (1954-2014).**

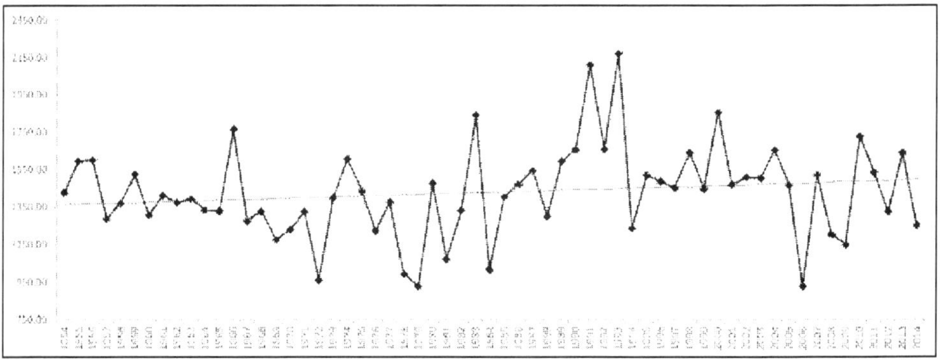

**Annual Rainfall of Imphal Region, Manipur (1954-2014).**

The state is projected to experience an increase in temperature above 1.7°C (Manipur State Action on Climate Change 2013). The projected increase in annual average temperatures for the southern districts are higher than the northern districts. the entire state of Manipur is projected to receive increased precipitation. The northern parts of the state are projected to experience an increase of ≥ 19 per cent of rainfall. This roughly correlates with observed trends over the last 30 years. The districts of Tamenglong and Senapati are projected to experience an increase in precipitation of ≥21 per cent; whereas, the southern districts also experience an increase in precipitation of ≥15 per cent. Further, an increase in the number of extreme rainfall (100 mm/day) conditions is projected for the state. Crop yields are projected to decrease by 10 per cent in 2030. An increasing trend has been observed in green house gas emission in Manipur from 1980 to 2005 (Jamir and De, 2013) in terms of $CO_2$ (2274 Gg/year rate of increase with 3.85 per cent compound growth rate), $CH_4$ (0.419 Gg/year rate of increase with 0.85 per cent CGR) and $N_2O$ (0.008 Gg/year rate of increase with 8.49 per cent CGR).

Hence, a high inter-annual variability in crop yields are projected due to increased frequency of extreme precipitation events. Incidence of pest and diseases as well as soil erosion with degradation is projected to increase. Increase in evaporation and runoff is projected which will decrease the soil moisture recharge. There is also a projection of scarcity of fresh drinking water. Upward movement of plant species is projected with rise in surface temperatures. Decreased microbial population is projected with increasing altitudes and loss of vegetation will pose potential threat for soil erosion. There is projection about loss of bio-diversity and extinction of rare/ threaten flora and fauna.

## Major Cropping Systems of Manipur

### a. Crop Sequences

1. Rice -fallow
2. Rice - Mustard
3. Rice –Pea

### b. Inter-Cropping

1. Potato + Mustard, Chilli + Bhindi
2. Potato + Broadbean, Chilli + Maize
3. Cabbage + Mustard, Ginger + Turmeric

### c. Mixed Cropping

1. Mustard and Pea
2. Sugarcane and Calocasia
3. Maize and Rice bean

### d. Catch Crops

1. Blackgram in rice field
2. Tomato in rice field
3. Rajmah in rice

### e. Crop Rotations

1. Not specified crop-relation is followed
2. Rice - Potato - Pea - Mustard
3. Maize - Potato – Beans

# 4

# Socio Economic and Farming System in Manipur

## Socio-Economic

The dictionary meaning of socio-economic is involving both socio and economic factors. In any society, the socio-economic factor plays a pertinent role. It influence and helps to understand the most resource management issues and to make sound resource management decisions. A survey of historical, economic and demographic activity helps to explain current geophysical and ecological conditions, such as hydrology, land use, water quality, and species diversity. The socio economic status includes general information of the village with the farmer specific information, the village specific information which includes the village name, block name, tahsil name, district name, institutions located in the village like post office, bank, and any other institution present in the village which delivers for the betterment of the society. The social status of the farmer includes variables like age, education, type of family, social participation, membership in farmers club or in any social organization, communication sources for the agri-allied activities, social contact with social leader, development agencies, village pradhan, district worker and any others. The Economic status includes the type of housing, size of holding, occupation, land operation includes total area, total cultivated area, orchard area, waste land, usar/unproductive land, land leased out, irrigation facilities like canal, well, tank, tube well, rain fed, farm power like bullock, buffalos, camel, tractors, farm implements like deshi plough, mould board plough, harrow, spade, hoes, seed drill, sickles, khurpi, sprayer, duster, oil engine, chaff cutter, cultivators and others, livestock includes cows, possession of livestock, their breeds, no of animal and their purpose like/milk. The material possession of the farmer includes the radio, tape recorder, television, chair, cycle, sofa, sewing machine etc.

### How to know the Socio Economic Profile of the District/Village/ Farmer?

The socio economic profile of the village or a farmer is known through data collection method. The data collection method includes Survey done through interview schedule and questionnaire, Participatory Rural Appraisal (PRA) techniques which incorporates number of techniques like focus group discussion, survey. The word "Survey" derived from sur + vey, sur means 'look' vey means over meaning by 'look over'. The survey is done through Interview schedule/Questionnaire.

"The survey is an organized attempt to analyse, interpretate the present status of a social institution, or group or area."

*– F.L. Whitney*

## Scope of Social Survey

According to Moser, Social survey helps to study the,

1. Demographic sphere - Family, Age, Education, Caste, Sex etc.
2. Social sphere- related to Social Pathology, Literacy, Crime, Alcoholism, Dowry etc.
3. Economic sphere- Poverty, Unemployment situation of farmers, Standard of living.
4. Cultural sphere- Attitude, Norms, Customs, Value, Recreation, Games.
5. Psychological sphere- Idea, Habit, Prejudice, Communication, Interaction etc.

## Procedure of Scientific Survey

### a. Problem to be Studied

1. Selection of problem
2. Nature of problem
3. Determination of objective
4. De-limitation of field of study
5. Deciding the unit of study
6. Formulation of hypothesis

### b. Preliminary Preparation

1. Preparatory study and experts opinion
2. Finance and budget
3. Time-limit and Schedule of work
4. Organisation and training

5. Type of survey
6. Method of survey
7. Test and method of measurement
8. Preparation of tools
9. Pre testing of tools- schedule/questionnaire
10. Final schedule/questionnaire

**c. Field Work and Classification of Data**

1. Collection of data
2. Classification of data
3. Tabulation of data
4. Checking of data
5. Supervision of organizer

**d. Generalization and Reporting**

1. Analysis of Data
2. Comparison and prediction
3. Reporting and display
4. Presentation of data- by graph, diagram etc.
5. Appendix- Sampling technique, tool used, test and measurement, references etc.

## Questionnaire and Interview Schedule

Questionnaire is the form which consists of set of questions wherein the respondents fill the answer by themselves. The respondent should be literate to fill the answer whereas, interview schedule is a technique of data collection wherein the interviewer personally visit the respondent, observes, ask the questions by themselves, record the observation and the facts told by the respondent. The respondent need not be literate in this case because the note making and the questions are clarified by the interviewer. The both the interview schedule and questionnaire takes lot of time and cost in terms of manpower utilization and the cost involved.

## Participatory Rural Appraisal

This is also a method of data collection wherein the data are collected in a rapid way by the team of people. PRA has its origin from RRA and has evolved and spread in early 1990's. It is said as a family of approaches and methods to enable local (rural or urban) people to express enhance, share and analyze their knowledge of life and conditions, to plan and to act (Robert chambers). Participatory Rural Appraisal (PRA) is widely practised nowadays in data collection. PRA brings together on the one hand, development needs defined by the community members and on the other, skills of Government, donor agencies and NGOs. It integrates traditional knowledge systems and external technical knowledge in the development process. PRA has many sources. The most direct is rapid rural appraisal (RRA) from which it has evolved. RRA itself began as a response in the late 1970s and early 1980s to the biased perceptions derived from rural development tourism (the brief rural visit by the urban-based professional) and the many defects and high costs of large-scale questionnaire surveys. The PRA techniques involves collection of basic information of the village, general transect, social map, agro-ecosystem map, resource map, venn diagram, seasonal calendar, wealth ranking, livelihood analysis/income and expenditure matrix, mobility diagram, time line, time trend, bio resource flow, technology table, pair wise ranking, matrix ranking, problem identification, ranking and prioritization, Indigenous Technical Knowledge (ITK), SWOT analysis. After getting the brief overview of the socio economic profile, the chapters of socio economic conditions portrays general picture of Manipur region with their district specific profile and moving towards village specific profile. The secondary data is collected for depicting the picture of Manipur Socio Economic Profile, District Socio Economic Profile. With respect to the data on Village specific profile, data is collected through interview schedule having a sample of 70 respondents from three villages of the three district mention below.

## Village Socio Economic Profile

Agriculture and allied activities (piggery, backyard poultry, fishery etc.,) form an important livelihood option for majority of both valley and hill region of Manipur. Both agriculture and allied activities contribute for the well being of the people. They are grown both for their consumption and for the market. Traditionally a farmer does farming and rear poultry, piggery, fishery etc., for meeting out their needs of the family. Basically, the north eastern people are meat lovers. They use varieties of animals for their consumption. Hence understanding the socio economic condition of the people is need of the time. Moreover, it is pertinent to document the socio economic condition of the farmers which would portray the wealth and status of the community in terms of agriculture and allied activities. It helps to plan various programmes and would help to devise new livelihood strategies for the community either for an individual/village as a whole.

## Methodology of the Study

To understand the socio economic condition as well as the integrated farming system scenario of the Manipur farmers, data has been collected with the help of well prepared interview schedule. Three hill districts have been chosen for the study. They are Churachandpur (Vazing village), Chandel (Chandel Khullen) and Ukhrul (Ramva Village) district was selected. In each district, 70 household was surveyed for both socio economic as well as agriculture related aspects.

This chapter will focus on

- ☆ Socio Economic condition of the farmer
- ☆ Traditional Farming system
- ☆ Economics of farming system
- ☆ Perception on Farm planning

- ☆ Agricultural problems faced by the farmers
- ☆ Thrust areas

## Socio Economic Condition of the Farmers

### Age

Age of the farmer reflects the percent of the people in the different age category. Nowadays, due to industrialisation, globalisation many farmers sell their own agricultural land for high price which is used as an real estate which intern used for buildings and other constructions. So the agricultural land is shrinking. Further many of the farmer's son and daughter have entered the mainstream and they are flourishing in other field. So young generation people are opting for better livelihood options rather than coming to do agriculture. Hence the category of people doing farming is the old age farmers. The following table portray the category of the farmers doing farming as a livelihood option. The data shows that, district wise, the percent of farmers falling in the category of less than 40 years is around 40 per cent. Similar trend also noticed in case of Chandel where 71 per cent of the farmers are in the age category of 40 years. The Ukhrul shows some difference as the farmers coming below 40 years is around 19 per cent. Manipur society in general has not developed compared to other states of India in terms of infrastructure *i.e.* Transport, power, water supply. Punitha (2012) in the report on the study on mass media, mention that 25 per cent of the hill farmers admit that there is no power supply in the remote areas of hill districts. Further, in valley region, 90 per cent of the farmers admit that there is less than 5 hours of power supply. The people of Manipur depend only on trucks to bring and take back the goods. The railway project is ongoing. Hence the young people who are in Manipur have no other option to diversify for jobs. Though there was in news that some big shots are buying agricultural land and using for non agricultural purpose. Organisations here also protesting against using for non agricultural purpose. Moreover, the real estate business which is very popular in other states also finds no picture in the state of Manipur. Hence it signifies that farmers really want to do agriculture in this region.

**Age of the Farmers**

| *Sl.No.* | *Age of Farmer* | *Churachandpur (N=70)* | | *Chandel (N=70)* | | *Ukhrul (N=70)* | |
|---|---|---|---|---|---|---|---|
| | | *No. of Farmers* | *Per cent* | *No. of Farmers* | *Per cent* | *No. of Farmers* | *Per cent* |
| 1 | <20 | nil | nil | 1 | 1.4 | nil | nil |
| 2 | 21-30 | 7 | 10 | 5 | 7.1 | 5 | 7.1 |
| 4 | 30-40 | 21 | 30 | 45 | 64.2 | 9 | 12.8 |
| 5 | 40-50 | 18 | 25.7 | 17 | 24.2 | 29 | 41.4 |
| 6 | >50 | 24 | 34.2 | 2 | 2.8 | 27 | 38.5 |

## Education of the Farmers

Educational status of the farmers is very important as it helps in number of ways. One among is that it helps to decide what mode of dissemination of the technology could be taken up. In a village, when majority are illiterate then any reading material will not give good mode of transfer of technology. The perusal of the educational status of the farmers in the following table tells that around 33 per cent of the farmers are literate in Churachandpur, 43 per cent of the farmers are literate in Chandel whereas 75 per cent of the farmers are literate in Ukhrul.

**Education of the Farmers**

| Sl.No. | Educational Status of the Farmer | Churachandpur | Per cent | Chandel | Per cent | Ukhrul | Per cent |
|---|---|---|---|---|---|---|---|
| 1. | Functionally Illiterate (No education) | 47 | 67 | 40 | 57 | 17 | 24.28 |
| 2. | Primary Education (5th class) | 2 | 2.85 | 5 | 7.14 | 17 | 24.28 |
| 3. | Middle Education (6-8 class) | 1 | 1.42 | 3 | 4.28 | 16 | 22.85 |
| 4. | Secondary Education (9th-12th class) | 14 | 20 | 15 | 21.42 | 15 | 21.42 |
| 5. | Collegiate | 6 | 8.57 | 7 | 10 | 5 | 7.14 |

## Family Size

Family size is nothing but the number of members present in the family. The family size is an advantage as it gives good strength in terms of domestic and farm management. It is evident that majority (82 per cent and 74 per cent in Ukhrul and Chandel respectively) possess the family size of more than 5. Moreover, based on the observation and the field survey, the number of children in the family in case of hills is more than 3. Even the poor people also have children more than 3. This is also one of the factors contributing for the bigger family size.

**Family Size of Surveyed Farmers of Manipur**

| Sl.No. | Churachandpur | | | Chandel | | | Ukhrul | | |
|---|---|---|---|---|---|---|---|---|---|
| | Family Size | No. of Farmers | Per cent | Family Size | No. of Farmers | Per cent | Family Size | No. of Farmers | Per cent |
| 1. | < 4 | 34 | 48.57 | < 4 | 18 | 25.71 | < 4 | 12 | 17.1 |
| 2. | > 5 | 36 | 51.42 | > 5 | 52 | 74.28 | > 5 | 58 | 82.8 |
| 3. | > 9 | nil | nil | nil | nil | nil | >9 | Nil | Nil |

## Occupation of the villagers

Village is the base for the development of any society. It is pertinent to know the village occupation. The following table portrays the occupation of the farmers in the village. It is observed that 51-57 per cent of the villagers are farmers. Less than 2 per cent are teachers. Around 4-10 per cent of the people are in police. Less than 4 per cent of the people are government employee. Around 5 to 15 per cent of them are carpenters.2-7 per cent are having an occupation as mason. Around 14 to 20 per cent are house wife doing small kitchen garden, rearing poultry birds for household consumption. Less than 2 per cent are pastor as the observation was made in the hilly districts. Hence from the above, it is known that the majority have farming as an occupation. Hence there is a lot of scope to uplift the farmers through scientific and technological intervention and to promote the allied activities in this region.

**Occupation of the Villagers**

| Sl.No. | Occupation | Churachandpur | | Chandel | | Ukhrul | |
|---|---|---|---|---|---|---|---|
| | | No. of Farmers | Per cent | No. of Farmers | Per cent | No. of Farmers | Per cent |
| 1. | Farmer | 36 | 51.4 | 40 | 57.14 | 39 | 55.7 |
| 2. | Housewife | 10 | 14.2 | 13 | 18.57 | 14 | 20 |
| 3. | Teacher | 1 | 1.4 | 2 | 2.85 | 1 | 1.42 |
| 4. | Police | 7 | 10.0 | 5 | 7.14 | 3 | 4.28 |
| 5. | Govt. employee | 1 | 1.4 | 2 | 2.85 | 3 | 4.28 |
| 6. | Carpenter | 11 | 15.7 | 5 | 7.14 | 4 | 5.71 |
| 7. | Mason | 2 | 2.8 | 3 | 4.28 | 5 | 7.14 |
| 8. | Pastor | 2 | 2.8 | Nil | – | 1 | 1.42 |

## Farm Size

Farm area is the area of the land owned by the farmer. The following table depicts that, the majority of the farmers in Churachandpur and Chandel districts respectively have a farm area less than 0.5 ha. In Ukhrul district, they possess farm area less than 1 ha. This means that land holding is very less for the hill farmers in Manipur. Hence, to maximise the profits from the small plot, farmers have to go for integrated farming system.

**Basic Information of Farm Size of Surveyed Districts of Churachandpur, Chandel and Ukhrul of Manipur**

| Sl.No. | Farm Size (ha) | Churachandpur (Vazing) N=70 | | Chandel (Chandel Khullen) N=70 | | Ukhrul (Ramva) N= 70 | |
|---|---|---|---|---|---|---|---|
| | | No. of Farmers | Per cent | No. of Farmers | Per cent | No. of Farmers | Per cent |
| 1. | < 0.5 | 46 | 65.7 | 36 | 51 | 22 | 31.4 |
| 2. | 0.6-1 | 1 | 1.4 | 33 | 47.1 | 37 | 52.8 |
| 3. | 1.1-1.5 | 2 | 2.8 | 1 | 1.4 | Nil | nil |
| 4. | 1.6-2 | 1 | 1.4 | nil | nil | 5 | 7.1 |
| 5. | > 2 | 3 | 4.3 | nil | nil | 1 | 1.4 |
| 6. | Landless | 17 | 24.2 | nil | nil | 5 | 7.1 |

## Farm Land (Own/Lease/Jhum Land)

There was a thought to see whether the farmer has cultivable land of his own or leased. Sometimes, farmer also possesses jhum land. Hence an idea has crept in mind to look into these aspects. It is noticed from the table given below that the owned land in Churachandpur district is 65.7 per cent whereas in case of Chandel and Ukhrul it is 98.5 and 90.2 per cent respectively. The farmer doesn't possess lease land in Ukhrul whereas, in case of Chandel and Churachandpur also it is meagre. It is also observed that the farmers from the Chandel and Ukhrul possess considerable per cent of jhum land *i.e.*, 57 per cent in Ukhrul and 74 per cent in Chandel.

**Farm Land Status in Surveyed Districts (Own/lease/jhum land)**

| *Sl.No.* | *Particulars* | *Churachandpur* | | *Chandel* | | *Ukhrul* | |
|---|---|---|---|---|---|---|---|
| | | *No. of Farmers* | *Per cent* | *No. of Farmers* | *Per cent* | *No. of Farmers* | *Per cent* |
| 1. | Own land | 46 | 65.7 | 69 | 98.5 | 65 | 90.2 |
| 2. | Lease land | 7 | 10 | 1 | 1.4 | Nil | nil |
| 3. | Jhum land | 3 | 4.2 | 52 | 74.2 | 40 | 57.1 |
| 4. | No land | 17 | 24.2 | nil | nil | 5 | 7.1 |

## Distance from the Market

This is the place where farmers normally travel to buy and sell goods. The goods may be groceries, vegetables etc. It is observed from the following table that the farmer approximately travel 6km to reach market in case of Churachandpur, less than 4 km in case of Chandel and 2km in case of Ukhrul district. The market does not give amenities to sell products in large quantity. But vegetables and meat products are sold in the market for daily consumption near to the village.

**Distance of Village from Market**

| *Churachandpur* | | *Chandel* | | *Ukhrul* | |
|---|---|---|---|---|---|
| Distance from the market | Per cent | Distance from the market | Per cent | Distance from the market | Per cent |
| 6 km (Churachandpur bazar) | 100 per cent | 1-4 km (Japhou bazar) | 100 per cent | 1-2 km (Ramva bazar) | 100 per cent |

## Farm Tools

Undoubtedly, the farm tools are very important for the cultivation of the crops. It is also important to note that whether the farmers have implements which are recent technology. The information is necessary to identify the resource base of the farmers. Based on this data, one could identify the type of technology already present with the farmers. And it helps to answer the question like what type of implements should be taken for the demonstration? Based on the number of tools possessed by the farmers it is ranked. The data in the following table reveals that the dao, spade, hammer, jumple, hammer, thapa are the implements most prominent among all the districts. But the advance technology like tractor, power tiller has found no place in the library of implements in these three districts.

**Farm Tools**

| Sl.No. | Name of Farm Tools | Churachandpur | | Chandel | | Ukhrul | |
|---|---|---|---|---|---|---|---|
| | | Number | Rank | Number | Rank | Number | Rank |
| 1. | Spade | 106 | I | 116 | II | 230 | I |
| 2. | Khurpi | 13 | | 2 | | 43 | |
| 3. | Sickle | 92 | II | 71 | IV | 175 | III |
| 4. | Rake | 15 | | 33 | | 49 | |
| 5. | Thapa | 15 | | 99 | III | 30 | |
| 6. | Dao | 26 | V | 127 | I | 180 | II |
| 7. | Load cart | 8 | | nil | | 10 | |
| 8. | Jumple | 47 | IV | 33 | | 55 | V |
| 9. | Hammer | 70 | III | 70 | V | 98 | IV |
| 10. | Power tiller | 1 | | nil | | nil | |

## Farming System Practised by Farmers

### Crops and Vegetables Grown by the Farmers

There are various crops and vegetables which are grown by the farmers. The table below shows that the farmers in the three districts done diversification of the crops in a limited area. In Churachandpur, the farmer grows variety of crops like spices, tuber crops, cereals, oilseeds, vegetables etc. Similarly, the farmers in Ukhrul also follows the same trend who grow variety of crops like spices, oilseeds, vegetables, cereals but more emphasis on vegetable crops. The cabbage, carrot, cauliflower made a good place in this district. In Churachandpur, rapeseed mustard, maize and other vegetables fetches higher income comparative to other crops. In case of Chandel, rice gives a good amount of income to the farmers. Other crops are grown not on commercial purpose. The excess vegetables are sold in the market. In Ukhrul, rice gives good amount of return followed by maize, cabbage, chow chow, turmeric.

### Livestock and Poultry

The piggery and poultry is one of the important animal components grown by the tribal people as a livelihood. The major pig feed the farmers use are, household waste, rice powder, banana stem, colocasia shoot. Among the all three hill district, for Ukhrul farmers, piggery and poultry both become the important part of livelihood. Similarly, for poultry, farmers use household waste, green vegetables, rice powder. In Ukhrul, farmers rear buffalo which is mainly used for consumption. Firstly, one has to understand that the tribals especially eat almost all type of meat which is a good source of protein. Sometimes, people fry and eat the honey bee. Dog is also used

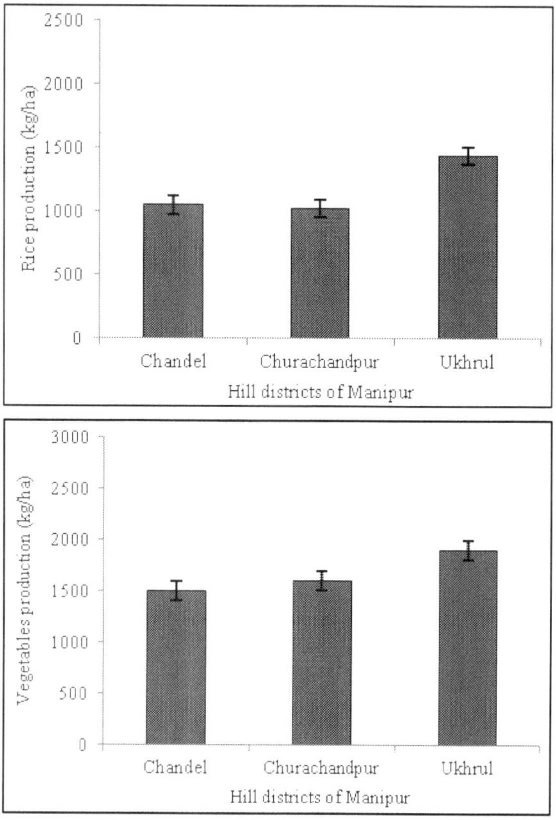


**Rice and Vegetable Production.**

sometimes for consumption. Hence, meat forms an important dietary constituent for the tribal people not only for Manipur but also for the whole North Eastern region.

**Livestock and Poultry**

| *Sl.No.* | *Components* | *TFS* |
|---|---|---|
| 1. | Piggery | Local Breed |
| 2. | Poultry | Local birds |
| 3. | Duckery | Local duck |
| 4. | Goatery | Undescript breed |
| 5. | Dairy (Cow, Buffalo) | Local breeds |
| 6. | Dog | Local breeds |

The major problem faced by the poor farmers in rearing poultry and piggery is that the feed. The quantity of feed which is consumed by both poultry and piggery is higher which is sometimes not affordable by the poor farmer. If the poor quality feed is used by the farmer then the pig/poultry will not gain weight, therein farmer incur loss. Hence government, if they provide subsidised feed for poultry and piggery it would really give cheer to the farmers and in turn will enhance the livelihood.

Among the price per kilogram of meat is concern, there is not much variation between districts. However, among the meat cost of goat, buffalo, cow meat goes up to Rs 250/kilogram. In some important festive season, the price may soar even higher than the above mention price. The milk is not consumed much by tribal people as compare to the meat part. It is consumed as a tea otherwise it is less consumed by the society.

**Feeding Management under Hill Districts of Manipur**

| *Sl.No.* | *Components* | *Churachandpur* | *Chandel* | *Ukhrul* |
|---|---|---|---|---|
| 1. | Piggery | a. Household waste | a. Banana stem | a. Banana stem |
| | | b. Rice bran | b. Rice bran | b. Paddy straw |
| | | c. Banana stem | c. Food waste | c. Colocasia shoot |
| | | d. Colocasia shoot | | d. Waste food |
| 2. | Poultry | a. Rice bran | a. Rice bran | a. Rice bran |
| | | b. Green vegetable | b. Household waste | b. Household waste |
| | | c. Household waste | | |
| 3. | Duckery | a. Household products | a. Household products | a. Household products |
| | | b. Rice bran | b. Rice bran | b. Rice bran |
| | | c. Feeding in water bodies | c. Feeding in water bodies | c. Feeding in water bodies |
| | | | d. Eichhornia | d. Eichhornia |
| 4. | Goatery | a. Green grass | a. Green grass | a. Green grass |
| | | b. Pasture | b. Pasture | b. Pasture |
| 5. | Dairy (Cow, | a. Paddy straw | a. Paddy straw | a. Paddy straw |
| | Buffalo) | b. Green grass | b. Green grass | b. Green grass |
| | | c. Pasture | c. Pasture | c. Pastured. Rice bran |
| 6. | Dog rearing | a. Household wastage | a. Household wastage | a. Household wastage |

**Quantity of Feed Used under TFS and IFS in Hill Districts of Manipur**

| *Sl.oN.* | *Components* | *Churachandpur (kg/unit/day)* | *Chandel (kg/unit/day)* | *Ukhrul (kg/unit/day)* |
|---|---|---|---|---|
| 1. | Piggery | 4.5 | 6.5 | 3.5 |
| 2. | Poultry | 0.44 | 0.45 | 0.35 |
| 3. | Duckery | 0.25 | 0.30 | 0.30 |
| 4. | Goatery | 4.0 | 3.0 | 3.0 |
| 5. | Dairy | 7.0 | 5.5 | 9.0 |
| 6. | Dog | 0.65 | 1.1 | 0.70 |

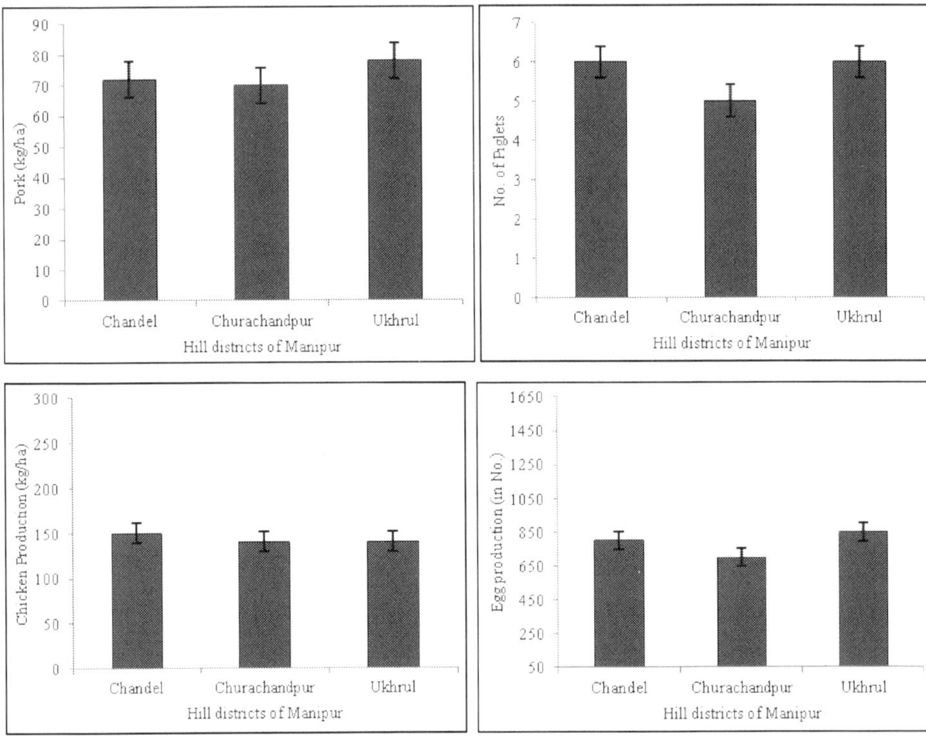


**Production Various Products in Chandel, Churachandpur and Ukhrul. Standard Error Bars Represent the Standard Error Value and Significant Level at P=0.05.**

## Fishery

Apart from livestock, fishery also plays an important role for the livelihood of the Manipur people. In Manipur, *phukri* locally called for farm pond found in most of the houses. Otherwise community ponds exist in many places. This is used as a multipurpose structure which is used for bathing, cleaning utensils, watering kitchen garden and so on. Some pond used for rearing fishes. The species they use are rohu, catla, mrigal etc. Further, for Manipur valley people, fish forms an important diet in day to day life. They use dry fish, fermented fish which is so popularly called as ngari as an important constituent in the diet of Manipur people. Further there is lot of scope for making fishery as an important source of livelihood by training the farmers in the area of fish seed production as there is a huge demand for fishes as well in composite fish culture. Apart from the above source of livelihood, people also earn money by cutting woods and selling them. Further selling of vegetables also plays an important role for improving the livelihood.

## Fruit Cultivation

North Eastern Region is said as an area rich in biodiversity. One could find on crops, vegetables, fruits grown by the farmers that the crops grown are also diverse. They grow many crops which may not be grown in a commercial way but few plants

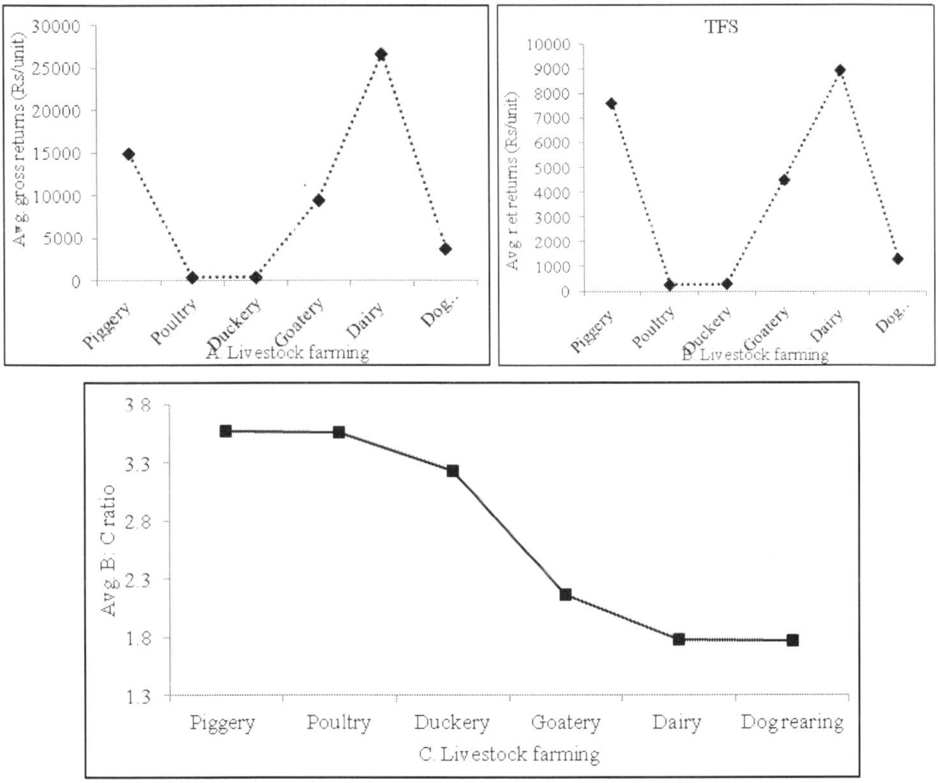


**Economics of Livestock (Average of three districts).**

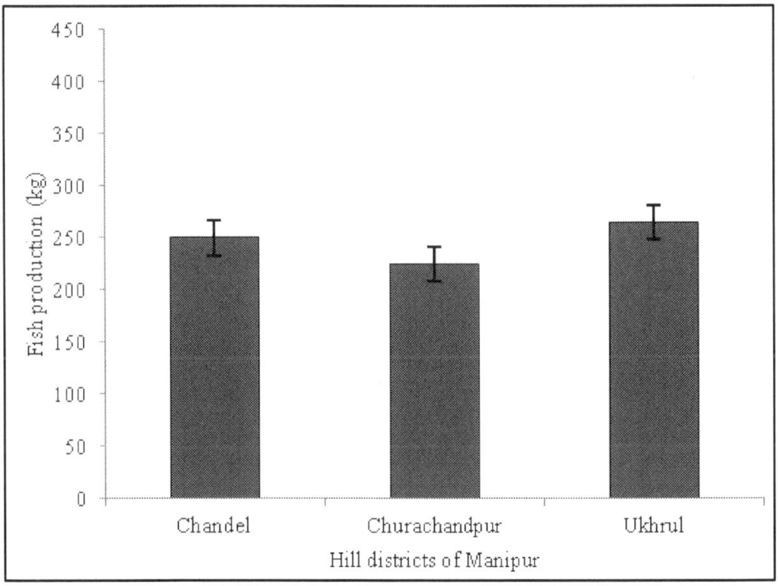


**Fish Production.**

**Fishery**

| *Sl.No.* | *Parameter* | *Churachandpur* | *Chandel* | *Ukhrul* |
|---|---|---|---|---|
| 1. | No. of ponds | 2 | 21 | 26 |
| 2. | Area (acre) | 0.25 | 9 | 8 |
| 3. | Seasonal/perennial | Perennial | Seasonal | Seasonal=7 Underground=2 = 17 |
| 4. | Name of the species | Silver carp, Rohu, Common carp | Silver carp, Rohu, Common carp | Silver carp, Rohu, Common carp |

atleast found a place in many household or a village. Among the district surveyed, the fruit crops grown by the Churachandpur are orange, lemon, pear, guava, papaya, jackfruit, aonla, mango, plum, passion fruit, mandarin which are grown as a plant occupied near the house grown for household purpose. The excess are sold to the market. In Chandel, kiwi occupies the prime position in fruit cultivation, followed by banana, orange, lemon and guava. In Ukhrul, like vegetables, varieties of fruits occupy the place in every household. The fruits namely, banana, papaya, plums, amla, guava, kiwi, pear gives a considerable earning to the farmers.

**Fruit Cultivation**

| *Churachandpur* | *Chandel* | *Ukhrul* |
|---|---|---|
| Pineapple | Banana | Lemon |
| Khasi mandarin | Guava | Plum |
| Lemon | Lemon | Guava |
| Pear | Khasi mandarin | Papaya |
| Guava | Passion fruit | Jackfruit |
| Papaya | Jackfruit | Banana |
| Jack fruit | | Khasi mandarin |
| Aonla | | Mango |
| Mango | | Pineapple |
| Plum | | Amla |
| Passion fruit | | Wild apple |

## Farm Produces and Farm Income

Farm income is the income farmers receive either through crop based activity, animal based activity. In Ukhrul district, after paddy, vegetable *i.e.*, cabbage followed by capsicum becomes the next income generating activity. In other districts, crops like turmeric, capsicum, tomato, ginger, onion, french bean forms an income generating source. In Manipur, the paddy is the major crop grown in the rabi season and grown as a rain fed crop. After kharif season, farmers are not going for rabi crop as intensively as kharif crop because of non availability of water for irrigating the crop. To combat the above situation, more and more emphasis is given to soil and water conservation

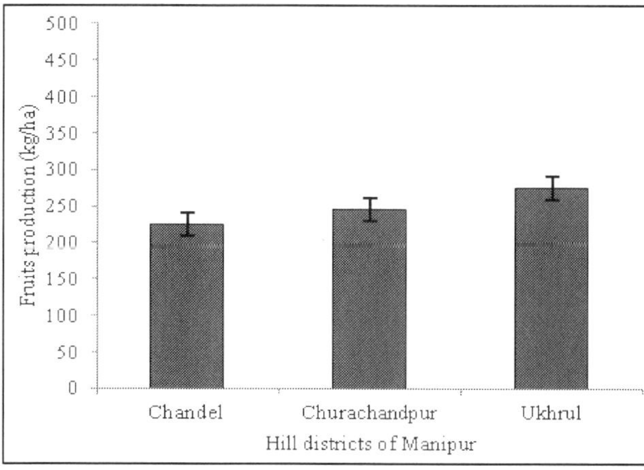


**Fruit Production.**

practices. The jalkund of 30,000 and 60,000 litre capacity water harvesting structure is promoted in both hills and valley region in Manipur through various projects like NICRA, TSP.

**Variable Costs of different Component of Farming Systems at Chandel, Churachandpur and Ukhrul District**

| *Sl.No.* | *Components* | *Variable Cost (Rs)* | | |
|---|---|---|---|---|
| | | *Chandel* | *Churachandpur* | *Ukhrul* |
| 1. | Paddy cultivation | 4,442.5±86.9 | 4,310.3±124.2 | 6,020.7±172.2 |
| 2. | Vegetable cultivation | 2,642.5±57.8 | 2,811.3±118.5 | 3,331.2±127.5 |
| 3. | Fruits cultivation | 1,501.7±30.3 | 1,543.5±57.0 | 1,543.3±106.5 |
| 4. | Piggery | 11,233.3±274.4 | 11,383.3±563.0 | 11,483.3±563.0 |
| 5. | Backyard poultry | 3,393.3±109.7 | 3,323.3±110.5 | 3,451.7±133.3 |
| 6. | Fishery | 3,983.3±274.4 | 3,643.3±127.1 | 4,116.7±116.7 |
| 7. | Water management | 150.0±5.8 | 150.0±5.8 | 199.3±13.9 |

**Gross Returns of different Component of Farming Systems at Chandel, Churachandpur and Ukhrul District**

| *Sl.No.* | *Components* | *Gross Return (Rs) from TFS* | | |
|---|---|---|---|---|
| | | *Chandel* | *Churachandpur* | *Ukhrul* |
| 1. | Paddy cultivation | 10,466.7±260.3 | 10,233.3±202.8 | 14,400.0±577.4 |
| 2. | Vegetable cultivation | 15,200.0±544.0 | 16,033.7±289.3 | 19,033.3±448.1 |
| 3. | Fruits cultivation | 4,550.0±217.9 | 4,900.0±144.3 | 5,500.0±230.9 |
| 4. | Piggery | 26,366.7±721.9 | 24,133.3±638.6 | 27,466.7±868.6 |
| 5. | Backyard poultry | 18,966.7±837.3 | 19,650.0±361.7 | 20,733.3±520.7 |
| 6. | Fishery | 24,900.0±838.7 | 22,400.0±608.3 | 26,366.7±696.0 |
| 7. | Water management | 350.0±6.90 | 320.0±5.01 | 450.0±8.3 |

**Net Returns of different Component of Farming Systems at Chandel, Churachandpur and Ukhrul District**

| Sl.No. | Components | Net Return (Rs) from TFS | | |
|---|---|---|---|---|
| | | Chandel | Churachandpur | Ukhrul |
| 1. | Paddy cultivation | 6,024.2±347.3 | 5,923.0±326.4 | 8,379.3±407.8 |
| 2. | Vegetable cultivation | 12,557.5±576.7 | 13,222.0±348.8 | 15,701.3±376.4 |
| 3. | Fruits cultivation | 3,048.3±187.8 | 3,356.5±200.8 | 3,956.7±335.7 |
| 4. | Piggery | 15,133.3±996.0 | 12,750.0±104.1 | 15,983.3±308.7 |
| 5. | Backyard poultry | 15,573.3±727.6 | 16,326.7±472.1 | 17,281.7±652.8 |
| 6. | Fishery | 20,916.7±1,112.2 | 18,756.7±481.5 | 22,250.0±579.5 |
| 7. | Water management | 200.0±15.8 | 170.0±15.8 | 250.7±116.3 |

**Benefit Cost Ratio (BCR) of different Component of Farming Systems at Chandel, Churachandpur and Ukhrul District**

| Sl.No. | Components | Net Return (Rs) from TFS | | |
|---|---|---|---|---|
| | | Chandel | Churachandpur | Ukhrul |
| 1. | Paddy cultivation | 2.4±0.1 | 2.7±0.1 | 2.4±0.1 |
| 2. | Vegetable cultivation | 5.75±0.3 | 6.40±0.3 | 5.70±0.3 |
| 3. | Fruits cultivation | 3.0±0.1 | 3.4±0.1 | 3.2±0.2 |
| 4. | Piggery | 2.3±0.1 | 3.0±0.1 | 2.1±0.1 |
| 5. | Backyard poultry | 5.6±0.1 | 6.0±0.1 | 5.9±0.3 |
| 6. | Fishery | 6.4±0.6 | 5.4±0.4 | 6.2±0.1 |
| 7. | Water management | 2.33±0.1 | 2.75±0.1 | 2.13±0.1 |

## Market Linked Expenditure

Farmers are expending in different sector for purchasing and full filling the monthly requirements from market.

**Market Linked Monthly Expenditure**

| Particulars | Chandel | | Churachandpur | | Ukhrul | |
|---|---|---|---|---|---|---|
| | Quantity | Amount (Rs) | Quantity | Amount (Rs) | Quantity | Amount (Rs) |
| Panparag | 80 unit | 500 | 60 | 350 | 150 | 750 |
| Gutka | – | – | – | – | 5 | 500 |
| Tamul | – | – | 2 | 100 | – | – |
| Chips | 46 unit | 930 | 40 | 400 | 80 | 1600 |
| Milk | 28 unit | 5560 | 30 | 600 | 25 | 1250 |
| Meat | 184 kg | 34,910 | 200 kg | 44000 | 160 kg | 32800 |
| Egg | 70 unit | 370 | 50 | 350 | 100 | 500 |

*Contd...*

**Contd...**

| Particulars | Chandel | | Churachandpur | | Ukhrul | |
|---|---|---|---|---|---|---|
| | Quantity | Amount (Rs) | Quantity | Amount (Rs) | Quantity | Amount (Rs) |
| Cigarette | 153 unit | 1430 | 100 | 1000 | 150 | 1500 |
| Khaini | 144 unit | 1420 | 120 | 2400 | 130 | 1300 |
| Tea | 95 unit | 1720 | 100 | 1000 | 130 | 1300 |
| Travel | – | 6000 | – | 6500 | – | 6000 |
| Medicine | – | 8000 | – | 8500 | – | 7500 |
| Education | – | 14,400 | – | 12000 | – | 18000 |

## Extend of Participation of Women in Decision Making

Decision making is one of the key aspects in any area. Women is basically described as a hand of doing multitask. However, their importance is neglected in much society. The survey was made to look into the participation of women in the decision making in the farming systems. Farming system as it comprises not only the crops but also the livestock, poultry, piggery, fishery, and so on. Basically the women inherently had the knowledge to do kitchen gardening, rear poultry birds, piggery may not be in a scientific way but they know various nuances underlie the cultivation of all the above. However, it is also necessary how far the woman capacity has been fully utilised in different decision making of the farming system is the question. The mean score was calculated and rank has been made. It is quite interesting to note from the data in the following table, which reveals that decision making is topped by the fertiliser to be used followed by means of irrigation and agriculture operations, crops to be sown, marketing/store of farm produce and sale of farm animals.

**Extend of Participation of Women in Decision Making in Farming Systems**

| Decision Making Area | Extant of Participation | | | | Mean Score | Rank |
|---|---|---|---|---|---|---|
| | Nil | Only Consulted | Opinion Considered | Final Decision | | |
| Crop to be sown | 12 | 123 | 68 | – | 0.55 | IV |
| Quality of seeds | 51 | 105 | 54 | – | 0.50 | VI |
| Fertilizer to be used | 15 | 114 | 90 | – | 0.61 | I |
| Means of irrigation | 22 | 98 | 88 | – | 0.58 | II |
| Agriculture operations | 31 | 79 | 100 | – | 0.58 | II |
| Labour to be hired | 57 | 73 | 80 | – | 0.52 | VII |
| Marketing/store of farm produce | 40 | 66 | 94 | – | 0.54 | V |
| Purchase and sale of farm animals | 53 | 68 | 89 | – | 0.54 | V |
| Purchase and sale of milch animals | 91 | 41 | 77 | – | 0.48 | VIII |
| Construction of farm buildings | 108 | 56 | 46 | – | 0.42 | X |
| Borrowing and repayment of farm loan | 101 | 49 | 60 | – | 0.45 | IX |

## Farmers Perception

**Farm Planning Consideration in Selection of Farm Activities (Weight age out of 10)**

| *Particulars* | *1* | *2* | *3* | *4* | *5* | *6* | *7* | *8* | *9* | *10* |
|---|---|---|---|---|---|---|---|---|---|---|
| *Churachandpur* | | | | | | | | | | |
| Food production for self- consumption | – | – | – | 3 | 19 | 2 | 3 | 1 | 1 | 41 |
| Combination of farm activity to supplement each other | 2 | 1 | 4 | 10 | 30 | 7 | 7 | 3 | 2 | 4 |
| Combination of farm Enterprises for maximum profit | – | – | 1 | 4 | 3 | 12 | 1 | – | 33 | 16 |
| Combination of farm enterprises for self]–consumption | – | – | 6 | 10 | 18 | 10 | 9 | 5 | 4 | 8 |
| As per available of infrastructure facilities | 1 | 2 | 13 | 11 | 11 | 14 | 12 | 6 | – | – |
| Availability of fund (own/barrowed) | 1 | 4 | 14 | 15 | 14 | 8 | 6 | 6 | 1 | 1 |
| Availability of reliable market | – | 2 | 3 | 9 | 25 | 16 | 12 | 2 | 1 | – |
| *Chandel* | | | | | | | | | | |
| Food production for self- consumption | 2 | – | – | 5 | 30 | 6 | 2 | 5 | 2 | 20 |
| Combination of farm activity to supplement each other | 1 | – | 10 | | 40 | | 3 | 3 | 3 | 10 |
| Combination of farm Enterprises for maximum profit | – | – | 2 | | 6 | | 2 | | 10 | 50 |
| Combination of farm enterprises for self]–consumption | – | 3 | – | 2 | 15 | 12 | 6 | 5 | 9 | 18 |
| As per available of infrastructure facilities | 2 | – | 9 | 20 | 18 | – | 7 | 9 | 5 | – |
| Availability of fund (own/barrowed) | 20 | 15 | 16 | 5 | 5 | – | 4 | 2 | 3 | – |
| Availability of reliable market | – | – | 3 | 7 | 25 | 10 | – | 5 | 20 | – |
| *Ukhrul* | | | | | | | | | | |
| Food production for self- consumption | – | – | 3 | 3 | 40 | 5 | 3 | 3 | 3 | 10 |
| Combination of farm activity to supplement each other | 2 | 2 | 4 | 4 | 30 | 8 | 7 | 2 | 5 | 6 |
| Combination of farm Enterprises for maximum profit | – | – | 1 | 2 | 18 | 9 | 8 | 2 | 15 | 15 |
| Combination of farm enterprises for self]–consumption | – | 5 | 2 | 6 | 10 | 6 | 9 | 12 | 15 | 5 |
| As per available of infrastructure facilities | 10 | 6 | 16 | 16 | 6 | 5 | 2 | 4 | 5 | – |
| Availability of fund (own/barrowed) | 20 | 10 | 7 | 10 | 5 | 9 | 2 | 5 | 1 | 1 |
| Availability of reliable market | – | – | – | 7 | 40 | 9 | 2 | 10 | 2 | – |

**Farmers' Perception on Farm Planning Consideration in Selection of Farm Activities**

| Particulars | Churachandpur | | Chandel | | Ukhrul | |
|---|---|---|---|---|---|---|
| | Mean Score | Rank | Mean Score | Rank | Mean Score | Rank |
| Food production for self-consumption | 2.9 | VI | 4.17 | V | 4.95 | III |
| Combination of farm activity to supplement each other | 5.58 | III | 5.24 | III | 5.25 | IV |
| Combination of farm enterprises for maximum profit | 2.85 | VII | 1.85 | VII | 3.68 | VII |
| Combination of farm enterprises for self]–consumption | 4.9 | V | 3.80 | VI | 4.28 | VI |
| As per available of infrastructure facilities | 6.01 | II | 5.78 | II | 4.72 | V |
| Availability of fund(own/barrowed) | 6.15 | I | 7.90 | I | 7.48 | I |
| Availability of reliable market | 5.58 | III | 4.68 | IV | 5.94 | II |

**Fertilizer used**

| Source of Nutrients | Churachandpur | | | | | |
|---|---|---|---|---|---|---|
| | Don't Like | Don't Like to Apply in Vegetable | Like to Apply in Rice | Like to Apply in all Crops | Mean Score | Rank |
| Urea | 22.4 unit | 21.12 | 13.44 | 13 | 2.24 | II |
| FYM | – | 8.6 | 28.38 | 32.68 | 3.32 | I |
| Poultry waste | 51 | 17 | 1.7 | – | 1.28 | V |
| MOP | 25 | 30 | 7.5 | 7.5 | 1.85 | IV |
| DAP | 10.35 | 27.6 | 27.6 | 4.14 | 1.96 | III |
| **Source of Nutrients** | **Chandel** | | | | | |
| | Don't Like | Don't Like to Apply in Vegetable | Like to Apply in Rice | Like to Apply in all Crops | Mean Score | Rank |
| Urea | 17.4 | 17.4 | 17.4 | 17.4 | 2.23 | II |
| FYM | – | 23.1 | 23.1 | 23.1 | 2.97 | I |
| Poultry waste | 42 | 7 | 7 | 14.0 | 1.90 | IV |
| MOP | 42 | 15 | 13 | – | 1.58 | V |
| DAP | 10 | 40 | 15 | 5 | 2.21 | III |
| **Source of Nutrients** | **Ukhrul** | | | | | |
| | Don't Like | Don't Like to Apply in Vegetable | Like to Apply in Rice | Like to Apply in all Crops | Mean Score | Rank |
| Urea | 15.75 | 9.45 | 25.2 | 18.9 | 2.65 | IV |
| FYM | – | 16.80 | 14 | 39.2 | 3.32 | I |
| Poultry waste | 14 | 7 | 21 | 28 | 2.9 | II |
| MOP | 7.7 | 15.4 | 23.1 | 23.1 | 2.86 | III |
| DAP | 9.6 | 16 | 32 | 12 | 2.65 | IV |

## Agricultural Problems Faced by the Farmers

It is evident from the following table that, in Churachandpur 97 per cent of farmers opines that they lack quality seed, followed by 87 per cent tells there is lack of seed treatment followed by moisture deficit in upland area followed by lack of seed of desired species, lack of irrigation facilities, rainfall during winter season is used for crop production. In Chandel also it is reported that lack of quality seed, followed by lack of seed of desired species, lack of irrigation facilities is the major problem faced by the farmer. In Ukhrul, the problem faced by the farmer are lack of seed treatment followed by lack of quality seed, lack of seed of desired species, knowledge of fertilizer application and so on. In Manipur, both in valley and hill region, farmers grow paddy as the major crop in the kharif season. But the quality seed is the question. Even the ICAR institute, is trying to produce quality seed for supply to the farmers. The state government has a greater role to play for seed production. Since the Manipur is landlocked as the transportation is either through air or truck or bus. It is very difficult to bring the quality seed material from outside. Based on the availability of the seed in market, the farmers are compelled to choose among the varieties available in the market. Further Manipur agriculture is rainfed agriculture. There is a no tube well or well irrigation practised by the farmer. The dams present are not able to cater to the needs of Manipur people. Hence there is no irrigation facility as mention by 71 per cent of the farmer in Chandel in Churachandpur.

**Agricultural problems**

| Sl.No. | Parameters | Churachandpur | |
|---|---|---|---|
| | | No. of Farmers | Per cent |
| 1. | Lack of Quality seed | 68 | 97.14 |
| 2. | Lack of seed treatment | 61 | 87.14 |
| 3. | Lack of seed of desired species | 54 | 77.14 |
| 4. | Rainfall during winter season is used for crop production | 47 | 67.14 |
| 5. | Much of the cultivated land left fallow | 34 | 48.57 |
| 6. | Lack of irrigation facilities | 50 | 71.42 |
| 7. | Moisture deficit on upland area | 59 | 84.28 |
| 8. | Excess moisture in valley land area | 45 | 64.28 |
| 9. | No technology for moisture management in low land area | 36 | 51.42 |
| 10. | No technology for moisture management on upland area | 36 | 51.42 |
| 11. | Lack of knowledge about reclaiming acid soil | 36 | 51.42 |
| 12. | Use of FYM/Compost at the of rate of t/h | 1 | 5.71 |
| 13. | Land is uneven or sloppy | 12 | 17.14 |
| 14. | Knowledge of fertilizer application | 40 | 57 |
| 15. | Use of recommended dose of fertilizer | 30 | 42.85 |
| 16. | Knowledge of irrigation management crops | 21 | 30 |
| 17. | Availability of credit facility | 12 | 32.85 |

*Contd...*

**Contd...**

| Sl.No. | Parameters | Churachandpur | |
|---|---|---|---|
| | | No. of Farmers | Per cent |
| 18. | Which fertilizer is likely accepted | | |
| | i. DAP | 10 | 14.28 |
| | ii. MOP | 1 | 1.28 |
| | iii. FYM | 17 | 24.28 |
| | iv. Potash | 16 | 22.85 |
| | v. Urea | 36 | 51.41 |
| | vi. Poultry waste | 1 | 1.28 |

| Sl.No. | Parameters | Chandel | |
|---|---|---|---|
| | | No. of Farmers | Per cent |
| 1. | Lack of Quality seed | 60 | 85 |
| 2. | Lack of seed treatment | 47 | 67 |
| 3. | Lack of seed of desired species | 50 | 71 |
| 4. | Rainfall during winter season is used for crop production | 30 | 42.8 |
| 5. | Much of the cultivated land left fallow | 40 | 57 |
| 6. | Lack of irrigation facilities | 50 | 71 |
| 7. | Moisture deficit on upland area | 40 | 57 |
| 8. | Excess moisture in valley land area | 10 | 14 |
| 9. | No technology for moisture management in low land area | 30 | 42.8 |
| 10. | No technology for moisture management on upland area | 20 | 28.5 |
| 11. | Lack of knowledge about reclaiming acid soil | 30 | 42.8 |
| 12. | Use of FYM/Compost at the of rate of t/h | 1 | 2.8 |
| 13. | Land is uneven or sloppy | 40 | 57 |
| 14. | Knowledge of fertilizer application | 20 | 28.5 |
| 15. | Use of recommended dose of fertilizer | 40 | 57.1 |
| 16. | Knowledge of irrigation management crops | 50 | 42.8 |
| 17. | Availability of credit facility | 7 | 10 |
| 18. | Which fertilizer is likely accepted | | |
| | i. DAP | 15 | 21.4 |
| | ii. MOP | 2 | 2.8 |
| | iii. FYM | 60 | 85 |
| | iv. Potash | 47 | 67 |
| | v. Urea | 50 | 71 |
| | vi. Poultry waste | 30 | 42.8 |

Contd...

**Contd...**

| Sl.No. | Parameters | Ukhrul | |
|---|---|---|---|
| | | No. of Farmers | Per cent |
| 1. | Lack of Quality seed | 50 | 71.4 |
| 2. | Lack of seed treatment | 60 | 85.7 |
| 3. | Lack of seed of desired species | 50 | 71.4 |
| 4. | Rainfall during winter season is used for crop production | 30 | 42.8 |
| 5. | Much of the cultivated land left fallow | 45 | 64.2 |
| 6. | Lack of irrigation facilities | 25 | 35.7 |
| 7. | Moisture deficit on upland area | 30 | 42.8 |
| 8. | Excess moisture in valley land area | 20 | 28.5 |
| 9. | No technology for moisture management in low land area | 20 | 28.5 |
| 10. | No technology for moisture management on upland area | 20 | 28.5 |
| 11. | Lack of knowledge about reclaiming acid soil | 20 | 28.5 |
| 12. | Use of FYM/Compost at the of rate of t/h | 2 | 5.7 |
| 13. | Land is uneven or sloppy | 40 | 57.1 |
| 14. | Knowledge of fertilizer application | 47 | 67.1 |
| 15. | Use of recommended dose of fertilizer | 30 | 42.8 |
| 16. | Knowledge of irrigation management crops | 20 | 28.5 |
| 17. | Availability of credit facility | 5 | 7.14 |
| 18. | Which fertilizer is likely accepted | | |
| | i. DAP | 20 | 28.5 |
| | ii. MOP | 25 | 35.7 |
| | iii. FYM | 50 | 71.4 |
| | iv. Potash | 60 | 85.7 |
| | v. Urea | 50 | 71.4 |
| | vi. Poultry waste | 30 | 42.8 |

## Constraints in Integrated High Value Farming System

The farmers in hilly region grow poultry or piggery either for family consumption or as a small scale business activity. It is also a fact that there is a high demand for meat products. The opinion of the farmers asked about not opting for high value integrated farming system which may give good income due to the existing demand in the market. The Churachandpur and Chandel hilly farmers opine that non availability of suitable infrastructure like sheds and houses was the first major constraint faced by the farmers. Lack of risk bearing ability, law and order problem, law of technical knowledge was the second constraint faced by the farmers of Chandel, Churachandpur, Ukhrul respectively. Lack of suitable market price also pointed as a constraint. Most of the farmers in hill region lead a subsistence agriculture. They don't have sufficient capital to invest in the hi value farming system. Frequent bandhs affect the movement of goods from one place to other hampers marketing. The hilly terrain makes the commodity more expensive and the suitable market price to the bulk product would not be attainable when the farmer goes for high value integrated farming system. Further, KVK in the hill districts should impart awareness and skill based demonstration on high value integrated farming system based on the need of the farmers. Hence, state government should bring favourable environment by providing easy access to loans, suitable infrastructure by constructing pucca roads and hi tech housing system and other need based infrastructure facilities, and providing subsidy for transportation of goods. The frequent bandhs hampering the social life should be curtailed so that the perishable commodity reaches market in time and in turn fetch good remuneration to the farmers.

| *Constraints* | *Churachandpur* | | *Chandel* | | *Ukhrul* | |
|---|---|---|---|---|---|---|
| | *Mean Score* | *Rank* | *Mean Score* | *Rank* | *Mean Score* | *Rank* |
| Non availability of suitable infrastructure like sheds and houses | 8.17 | I | 8 | I | 7.11 | III |
| Scarcity of family labour due to involvement in non farming activity | 4.71 | VII | 5.25 | VI | 4.91 | VII |
| No attractive remunerative | 4.97 | VI | 7.04 | III | 5.41 | VI |
| Irregular supply of own fund/margin Money Required for Diversifying farm activities | 5.61 | V | 4.5 | VII | 4.6 | VIII |
| Lack of technical knowledge | 5.97 | IV | 4.75 | VI | 7.12 | II |
| Social/caste factor in adopting like piggery, goatry etc. | 9.62 | 1X | 9.71 | IX | 9.57 | IX |
| Lack of risk bearing ability | 7.25 | II | 5.45 | V | 6.45 | V |
| Lack of guaranteed/suitable market price | 6.21 | III | 6.27 | IV | 6.58 | IV |
| Law and order problem | 3.22 | VIII | 7.81 | II | 8.14 | I |

## Thrust Areas Identified for Manipur

1. Development of location specific farming system models.
2. Introduction of high yielding varieties of cereals, pulses and oilseeds.
3. Soil and seed treatment.
4. Proper nutrient management.
5. Proper crop management and control of weeds.
6. Introduction of high yielding varieties of fruits and vegetables.
7. Improved production technology of fruits and vegetables.
8. Introduction of biofertilizers *e.g. Rhizobium, Azotobacter, Azospirillum,* Blue green algae, *Azolla* for nutrient management.
9. Proper design of terrace, water harvesting and designing diversions for irrigation and drainage for
    a. Proper management of watershed area.

10. Use of suitable plant protection measures against pest and diseases of crops.
11. Introduction of improved storage structure for cereals, pulses and finished products
12. Introduction of quality livestock germplasm.
13. Vaccination and health coverage measures of pig and poultry.
14. Deforming of livestock at regular intervals.
15. Development of capabilities of rural youth and women in the field of fruits and vegetables processing and value addition
16. Introduction of common carps and other exotic carps in paddy cum fish farming and fish ponds
17. Strengthening integrated farming system approach with advanced scientific inputs.

## Focus for Agricultural Development in Manipur

1. To increase the productivity of land through situation based Integrated farming system/agro forestry intervention.
2. To identify the important problems of that area as per the need of the farmers and prioritization of the identified problems as per their importance.
3. Planning and conducting on farm trial to assess the technologies in farmer's situation in the state and refine it, if required.
4. Demonstrating new and improved technologies to the farmers as well as to the extension agencies directly in the farmer's field with their active participation for promotion and large scale adoption.
5. Planning and conducting production oriented need based long and shorts duration on campus and off campus training.
6. Providing training and information about latest developments in agricultural technologies to extension personnel to orient and update their knowledge.
7. Developing and organizing non-formal education programs by way of field days, farm visits, farmers fair, radio talk, film shows etc.
8. Organizing Farm Science Clubs in Rural Schools and Villages to induce liking and interest for Agricultural and Allied sciences in younger generation.
9. Developing and promoting small village organizations like Self Help Groups (SHGs), Farmer Interest Groups (FIGs), Kisan Clubs etc.
10. Developing and maintaining farms and Demonstration Units on scientific lines to facilitate work experience to the trainees and also to put a show case of latest technical know how.
11. Imparting some basic education to rural illiterates and school drop-outs in order to make them not only good farmers but also better citizen

12. Providing added training facilities in the areas of rural home science and nutritional education for community.
13. Collecting feedback from the farmers and extension agencies and communicating these messages to research scientists for modification of technology.

# 5

# Strategies for Development

## Module I (Integrated Farming System)

All over the North East Region, farmers work hard but do not make money, especially small farmers because there is very little left after they pay for all inputs (seeds, livestock breeds, fertilizers, pesticides, energy, feed, labour, etc.). The emergence of Integrated Farming Systems (IFS) has enabled us to develop a framework for an alternative development model to improve the feasibility of small sized farming operations in relation to larger ones (Ravisankar *et al.*, 2006). Integrated farming system is a commonly and broadly used word to explain a more integrated approach to farming as compared to monoculture approaches. It refers to agricultural systems that integrate livestock and crop production or integrate fish and livestock and may sometimes be known as Integrated Bio systems. In this system an inter-related set of enterprises used so that the "waste" from one component becomes an input for another part of the system, which reduces cost and improves production and/or income. IFS works as a system of systems. IFS ensure that wastes from one form of agriculture become a resource for another form. Since it utilizes wastes as resources, we not only eliminate wastes but we also ensure overall increase in productivity for the whole agricultural systems. We avoid the environmental impacts caused by wastes from intensive activities such as pig farming.

These enterprises not only supplement the income of the farmers but also help in increasing the family labour employment.

1. The integrated farming system approach introduces a change in the farming techniques for maximum production in the cropping pattern and takes care of optimal utilization of resources.

2. The farm wastes are better recycled for productive purposes in the integrated system.
3. A judicious mix of agricultural enterprises like dairy, poultry, piggery, fishery, sericulture etc. suited to the given agro-climatic conditions and socio-economic status of the farmers would bring prosperity in the farming.

## IFS: Concepts and Significance

Farming and agriculture are respectively defined as the practice of cultivating the land or raising stock and the production of agricultural goods through the growing of plants and the raising of domesticated animals.

In the least development regions (NEH Region) frantic efforts should be made through research and enlightenment campaigns to encourage farming and thus ensure significant increases in agricultural production in order to feed and sustain the population that is increasing at geometric proportion. However, the increases recording have often fluctuating considerably due to improper coordination of measures that support increased productivity (Othman, 2006). The sustenance of increased productivity must emphasize on the development of strategies aimed at maintaining improved yields without depleting natural resources or destabilizing the environment. Such strategies abound in IFS. Integrated farming (or integrated agriculture) is a commonly and broadly used word to explain a more integrated approach to farming as compared to existing monoculture approaches. It refers to agricultural systems that integrated livestock and crop production. Integrated farming system has revolutionized conventional farming of livestock, aquaculture, horticulture, agro-industry and allied activities (Chan, 2006). It could be crop-fish integration, livestock-fish integration, crop-fish-livestock integration or combinations of crop, livestock, fish and other enterprises (Tokrishna, 2006). Integrated farming system can be complete integration encompassing crops, livestock, fisheries, processing and biogas units (Igbinnosa and Okporie, 2007) or partial integration involving different combinations of the later units. Whichever is the case, IFS can remove all the farming constraints (such as shortage and high cost of inputs and environmental pollution) by not only solving most of the existing economic and even ecological problems, but also provide the needed means of production such as fuel, fertilizer and feed, besides increasing productivity many-fold. It can turn all the existing disastrous farming systems; especially in the poor countries into economically viable and ecologically balanced systems that will not only alleviate poverty, but can even eradicate it completely [Chan, 2006].

## Scope of Farming System

Farming enterprises include crop, livestock, poultry, fish, tree crops, plantation crops, etc. A combination of one or more enterprises with cropping, when carefully chosen, planned and executed, gives greater dividends than a single enterprise, especially for small and marginal farmers. Farm as a unit is to be considered and planned for effective integration of the enterprises to be combined with crop production activity. Integration of farm enterprises to be combined on many factors such as:

- ☆ Soil and climatic features of the selected area.

- ☆ Availability of resources, land, labour and capital.
- ☆ Present level of utilization of resources.
- ☆ Economics of proposed integrated farming system.
- ☆ Managerial skill of the farmer

In the context of North East Regions, there are a number of situations and conditions that can be alleviated by an Integrated Farming System. The following situations are ideal for the introduction of IFS:

- ☆ The farmer wishes to improve the soil quality
- ☆ The farm household is struggling to buy food or below the poverty line
- ☆ Water is stored on-farm in ponds or river-charged overflow areas
- ☆ Fertilizers are expensive or the recommended blend is unavailable
- ☆ Soil acidity problems
- ☆ The farmer is seeking to maximize profits on existing holding
- ☆ The farm is being eroded by wind or water
- ☆ The farmer is looking to reduce chemical control methods
- ☆ The farmer wants to reduce pollution or waste disposal costs

The practitioner can use IFS to:

- ☆ Improve productivity
- ☆ Regulate nutrient and material flows
- ☆ Increase on-farm biodiversity
- ☆ Limit disease
- ☆ Reduce the smell of some livestock operations

## Integration of Choices

In agriculture, crop production is the main activity. The income obtained from crops may hardly be sufficient to sustain the farm family throughout the year. Assured regular cash flow is possible when the crop is combined with other enterprises. Judicious combination of enterprises, keeping in view of the environmental conditions of a locality will pay greater dividends. At the same time, it will also promote effective recycling of residues/wastes.

## Choice of Enterprises

Livestock is the best complimentary enterprise with cropping. Installation of composting and vermicomposting unit in crop-livestock system will make use of the wastes.

## Components of IFS

The components of IFS include crops, fish farming, poultry, pigs, cattle and goat, fodder production, kitchen gardening. The feeds derived from "alternative" crops (maize, paddy and leaves of trees and grasses) require "alternative" farming systems.

These are on small-scale and are highly productive. These are diversified and integrated and the role of animals in these systems is synergistic rather than as primary producers. Emphasis is on "small" livestock. External inputs can be minimized through waste recycling, and growing of nitrogen-fixing and pest-resistant plants in the farming system. The model provided below is integrated farming system that combines cash crop/tree crop/livestock mix which is functionally sustainable and profitable. The model shows the basic components and interaction used in actual integrated farming systems. IFS models could be developed based on existing production system using concepts and components. The farm has subsystems:

- ✰ Crop production including vegetables
- ✰ Fodder production
- ✰ Pig production
- ✰ Poultry production (Dual purpose)
- ✰ Cattle production (Dairy/Beef)
- ✰ Feed conservation and management
- ✰ Compost/Humus production

Each can stand-alone system but with the addition of composting and feed management unit and introduction of earthworms as a means of producing organic fertilizer the farm becomes one large efficient unit.

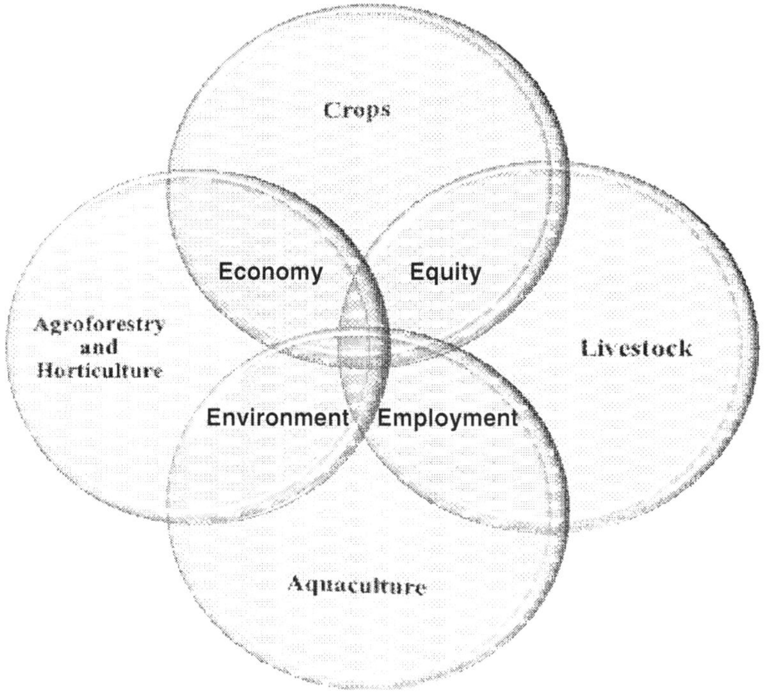


**Integration of Crops, Livestock, Aquaculture, Agroforestry and Horticulture within a Production System while Addressing the Social, Economical and Environmental Concerned for Sustainable Ecosystem.**

# IFS Module

- ☆ **Maize/Beans/Cash crops:** To produce adequate maize and beans (Red gram, cowpea, black gram/greengram) for farm use as well as for sale. The maize will feed the family and poultry, and along with residues from the root crops and fodder tree leaves can fatten the pigs.
- ☆ **Fishery:** The pond can be dug to serve as a water catchment area. This could supply irrigation water in the dry season. Fish culture at the start of

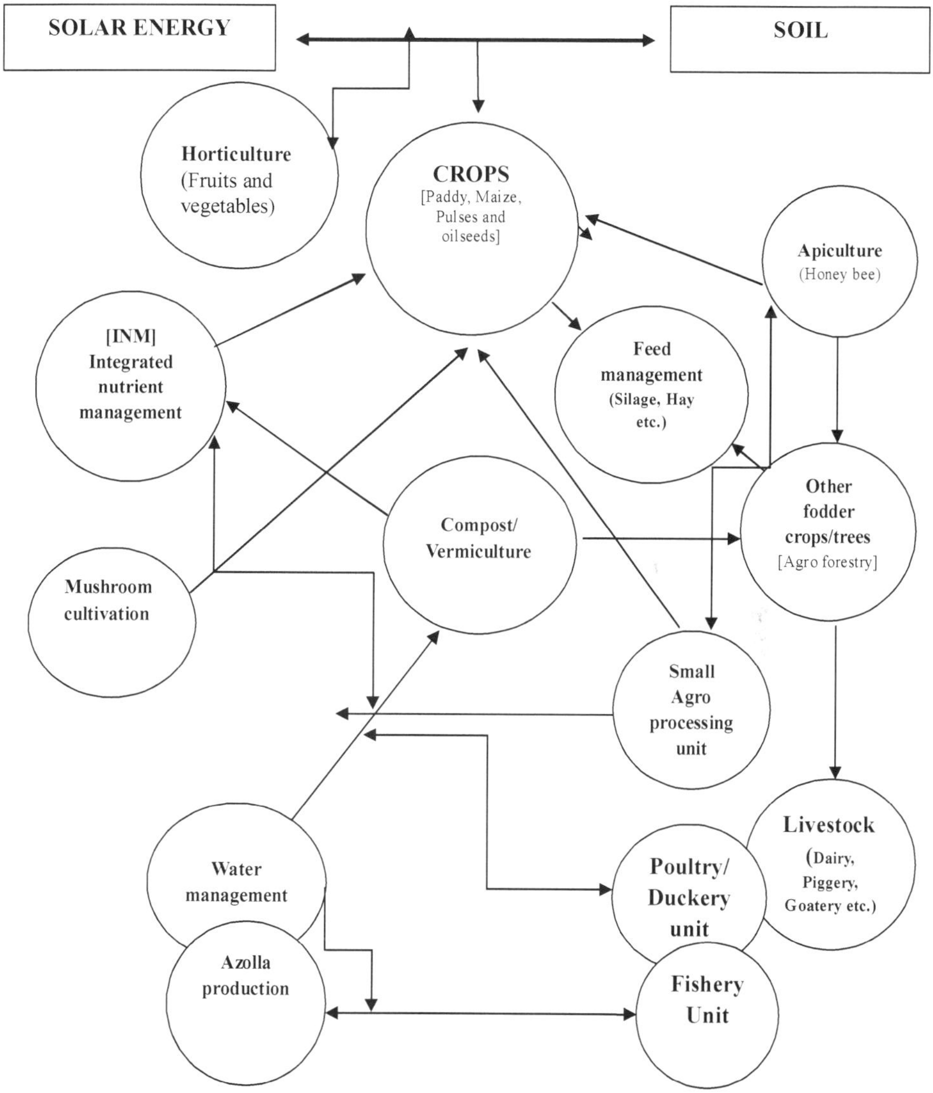


**Model of Integrated Farming System Developed by ICAR RC for NEH Region, Manipur Centre, Imphal (This model is being implemented at Chandel (Monsangpantha), Churachandpur (Moldennom), Imphal West (Kangchup), Tamenglong (Noney Peaceland), Ukhrul (Patsai) and Senapati (Purul) district) at farmers field.**

the wet season with harvesting at the end of the dry season when the water reserve should be reduced. Commercial fishery production should be based on commercially prepared feed use to fatten 4,000 to 6,000 fishes in acre pond over 5 to 6.

- **Horticultural crops:** Horticultural crops planted to a mixture of banana, citrus, pomegranate, mango, yam, passion fruit, papaya and vegetables for sale and use in the home and rejects could be fed to fattening pigs.
- **Fodder crops:** All areas planted to permanent crops would be separated by fodder trees like Leucaena, grasses and hedges. These would be sources of feed for livestock. Tree been should also be inter-planted in the hedges to supply additional income.
- **Poultry:** The family should maintain a dual purpose flock of adult layers. Feed consumption should be full fill from corn, and supplementary feeds and other leguminous leaves. The two local hens should be used mainly as brood to set the eggs. 24 eggs approximately should be set every 3 months. This would allow farm family to produce eggs and bird while maintaining the flock at no cost.
- **Piggery:** The farmer can replace commercial feed with various crops grown on the farm. From weaning to market weight a pig can be fattened on 10 Leucaena trees, yam and soybean protein and other pulses and farm produce. These could also be fed on well balanced silage.
- **Small agro processing unit:** Most of the farm produce like horticultural crops (Mango: pickles, Amchur; Banana: Banana chips; Citrus: pickles, Passion fruits: Juice etc.) can be use as a processed products. Those have higher economical value than the raw product.
- **Mushroom cultivation:** Paddy straw can be use for the cultivation of paddy straw mushroom at low cost level.
- **Apiculture:** Apiculture is a high value based components of farming system. The honey box can be keep with mustard cultivation and combine with agroforestry system.
- **Compost/Vermicomposting:** These will provide organic fertilizer by utilizing farm wastes.

## Goals of IFS

The Goals of Integrated Farming Systems (IFS) are to:

- provide a steady and stable income rejuvenation/amelioration of the system's productivity and
- Achieve agro-ecological equilibrium through the reduction in the build-up of pests and diseases, through natural cropping system management and the reduction in the use of chemicals (in-organic fertilizers and pesticides).

# Strategies for Higher Productivity and Production

## Adoption of Better Cropping Pattern which will Match According to Soil Condition, Climate and Quality of Water

In Manipur, less geographical area is utilized for cultivation of agricultural crops, however, there is a little scope to increase the area under cultivation of agricultural crops. About 98 per cent cultivated area is rainfed. This indicates that there is scope for diversification of food crops based on cropping systems approach under specific management.

The agricultural practices in the region are broadly of two distinct types *viz.*, settled farming practiced in the plains, valley/foot hills and terraced slopes and shifting cultivation in the hill slopes. Depending upon the system of farming, food habits and climatic conditions, several crops are grown in the region.

**Existing System in Manipur**

| *Valley Region* | *Hill Region* |
|---|---|
| Farm ponds near home yard or within paddy field, fish + livestock + kitchen garden + vegetables + rice,use of farmyard manure, compost, life saving irrigation with harvested water in pond | Maize, Groundnut, Soybean, Upland rice, Vegetables etc. |
| **Cropping intensity (137 per cent)** | |
| **Alternate and viable system for Manipur** | |
| Farming system (with all components)<br>**Kharif:** Rice, horticultural crops vegetables and fruits<br>**Rabi:** Mustard, Lentil, rabi maize, mung, Lathyrus, Pea, vegetables. | **Kharif**: Maize, Groundnut, Soybean, Upland rice, Vegetables etc.<br>**Rabi:** Mustard, Lentil, rabi maize, urd bean, mung bean, Lathyrus, Pea, vegetables (radish, potato). |
| **Cropping intensity (> 200 per cent)** | |

## Selection of Drought Avoidance Crop and Varieties/Species with High Water Use Efficiency Under Rainfed Conditions

| *Crops* | *Varieties* | *Yield (q/ha)* | *Salient Features* |
|---|---|---|---|
| | | | **Cereals** |
| **Rice** | RC Maniphou-4 | 45-50 | Semi-dwarf variety, panicle compact and fully exerted (19-22cm), grains medium fine, non aromatic. Recommended for growing in early *kharif* in valley areas of Manipur, photo-insensitive, it matures in about 115 days and escapes from blast reaction. |
| | RC Maniphou-5 | 45-50 | Semi-dwarf variety, panicle compact and fully exerted (19-22cm), grains medium fine, non aromatic. Recommended for growing in early *kharif* in valley areas of Manipur, photo-insensitive, it matures in about 115 days and escapes from blast reaction. |

*Contd...*

**Contd...**

| Crops | Varieties | Yield (q/ha) | Salient Features |
|---|---|---|---|
| | RC Maniphou-6 | 40-45 | Tall variety, panicle long and well exerted panicle. Recommended for growing in irrigated wetland of valley areas of Manipur during early *kharif* and raifed uplands of *kharif*, photo-insensitive, duration from seedling to 50 per cent flowering is recorded 85-90 days in *kharif* and 100 - 105 days in early *kharif*. |
| | RC Maniphou-7 | 60.0 | Semi-dwarf variety well exerted panicle, grains long slender. Recommended for irrigated/waterlogged(5-15cm) lowland with medium sowing and early to late planting in Manipur during *kharif*, slightly photo-sensitive. |
| | RC Maniphou-10 | 57-65 | Semi dwarf, glutinous rice variety suitable for valley areas, moderately resistant to gall midge, stem borer, blast and BLB. |
| | RC Maniphou-11 | 60.0 | A medium duration variety (135 days), plant height: 100-105 cm long, slender grain, soft, cooked rice, intermediate in amylase content, resistance to blast and moderately resistant to Brown Plant Hopper. |
| | RC Maniphou-12 | 40-45 | The variety takes 75 days to reach 50 per cent flowering and hence, was found suitable for pre-kharif/early kharif/ main kharif (contingency variety). It has desirable soft cooking (low amylose content-11.70 per cent) quality characters preferred by the people of north eastern hill region (NEH). So far as low amylose content is concerned it has been reported in traditional land races. It is a first evolved culture having very low amylose content. It has several merits over traditional long duration land races as it matures in about 95-105 days and it will be very much suitable for different cropping systems. |
| | Bhalum-1 | 35-38 | A medium duration variety, highly resistant to blast, suitable for uplands in mid altitude areas. |
| | Bhalum-3 | 35-38 | Recommended for rainfed uplands during main *kharif*, also suitable for Jhum cultivation and gives substantial yield. Resistant to blast and is non-lodging |
| | Bhalum-4 | 35-38 | Recommended for rainfed uplands during main *kharif*, also suitable for Jhum cultivation and gives substantial yield. Resistant to blast and is non-lodging. |
| | Megha SA-2 | 32-38 | Recommanded for mid altitude rainfed areas upto 1000 above msl. Resistant to blast and is non lodging. Maintains moderate aroma when grown in hills. Its duration is 149-155 days |
| | Leimaphou/ (KD-2-6-3) | 50 | Semi- dwarf, SL grains, suitable for late planted situations, moderately tolerant to major insect pests and diseases of Manipur |
| | Pariphou | 45-50 | Semi-dwarf, mildly scented, SL grains, drought tolerant, early maturing rice, moderately tolerant to major insect pests and diseases occurring in rice. |

*Contd...*

***Contd...***

| *Crops* | *Varieties* | *Yield (q/ha)* | *Salient Features* |
|---|---|---|---|
| | Erima phou (KD-7-9-20) | 40-45 | Slender long grains, with kneeing ability, tall rice, moderately tolerant to major insect pests and diseases of Manipur. Suited for semi deep to deep water condition. |
| | Akutphou | 40-46 | Semi-tall, SL grains, reddish husk with white rice, moderately tolerant to major insect pests and diseases of Manipur. |
| **Maize** | Ganga-9 | – | Yellow semi-flint type, matures in 100-110 days |
| | Him-129 | 40 | Yellow, semi-flint type, tolerance to MLB, TLB, ESR |
| | Ganga-11 | 50 | Yellow semi-flint type, late maturity duration, tolerance against DM. TLB |
| | Pratap Makka-4 | 45 | White, semi-flint, tall type and it shows moderately resistance to stem borer |
| | Hunius | – | Early, orange semi-flint type matures in 80-85 days |
| | VL Makka-42 | 25 | Orange, flint type, extra-early maturity, resistance against Stem borer |
| | Vivek-21 | 50 | Yellow-semi flint type, extra-early maturity duration type, tolerance against TLB. |
| | Pusa Composite-3 | 35 | Yellow-orange, flint type, medium maturity duration, tolerance to stalk borer, resistance to lodging and tolerance to moisture |
| | Pusa Composite-4 | 36 | Yellow-flint type, early maturity type, tolerance to stalk borer, lodging resistance and tolerance against water stress. |
| | Vivek-25 | 45 | Yellow, semi-dent type, extra early maturity type, tolerance against TLB |
| | Prakash | 48 | Yellow, flint type, Resistance to lodging, shattering, MLB, BSDM resistance |
| | HQPM-1 | 45 | Quality protein maize type, matures in about 110-120 days, tolerance to frost/cold, resistance to MLB and common rust |
| | Vivek QPM-9 | 49 | Yellow- dent, QPM type, extra early maturity, performed better at low Nitrogen |
| | PMH-1, PMH-4 | 40-45 | Medium duration hybrid good for hilly areas |
| | | | **Pulses** |
| **Blackgram** | T-9 | 8-10 | Suitable for rice fallow cropping system, maturity duration 65-70 days |
| | ADT-5 | 8.0 | Selection from germplasm of Kanpur, tolerant to MYMV, matures in about 65-70 days |
| | TAU-2 | 10.0 | A derivatives of T-9XU-196, matures in 70 days |
| | TPU-4 | 74.0 | Selection from Kanpur germplasm, tolerant to MYMV |
| | Jawahar Urid-2 | 13.0 | Tolerant to Macrophomina, CLS and MYMV |
| | Pant U-35 | 10.8 | Hairy pods, resistant to MYMV, matures in about 75-80 days |

*Contd...*

***Contd...***

| *Crops* | *Varieties* | *Yield (q/ha)* | *Salient Features* |
|---|---|---|---|
| **Green gram** | AAU-34 | 10.0 | Tolerance to MYMV, matures in 60-65 days (early type) |
| | Pant Moong-2 | 10-12 | Plant semi erect, shining green, seed weight 3.5-4.0 g/ 100 seeds, matures in 60-65 days |
| | Pant Moong-4 | 12-15 | Plant erect, seed dull green, seed weight 3.0 g/100 seeds, multiple disease resistant, matures in about 65-70 days |
| | PDM-54 | 9.11 | Plant erect, bushy, seeds shining green. Rectangular bold (4g/100 seeds), matures in 65 days |
| | PS-16 | | |
| | Pusa Baisakhi | | Dwarf early type matures in 60-65 days suitable for summer/spring crops |
| **Rice bean** | MNPL-1 | 13.09 | Ranked 1st preference among farmers of Manipur due to higher yield, bold seed and good taste. |
| | MNPL-2 | 8.41 | Normal duration variety having bold seed, good in taste and medium plant height. |
| | RCRB-1-6 | 12.90 | High yielding, early maturity type |
| | RBL-1 | 8.72 | Developed by Punjab and released for north west and north east regions, spreading variety with lateral inter wining branches with vigorous growth and profuse podding. Its matures in about 125-135 days |
| | RBL-6 | 6.06 | Developed by Punjab and released for north west and north east regions, spreading variety with lateral inter wining branches with vigorous growth and profuse podding. Its matures in about 125-135 days |
| | BRS-2 | 7.00 | Higher yield and attractive seed colour |
| | PRR-1 | 9.61 | High yielding variety |
| **Field Pea** | Rachna | 20-25 | Resistant to powdery mildew, dark green foliage matures in about 120-125 days |
| | HPP-4 | 26 | Plant dwarf, leafless, resistant to powdery mildew, matures in about 145 days |
| | Malviya Matar-2 | 21 | Plant erect, tall, creamy white bold seed, matures in about 100-140 days |
| | Uttara | 22.3 | Dwarf type plants, resitant to powdery mildew, matures in about 128 days |
| | Alankar(DMR-7) | 23.5 | Resistant to powdery mildew, matures in about 135 days |
| | Prakash | 18-22 | Good for timely sown |
| **Chickpea** | Phule G -95418 | 18-20 | Resistant to wilt, it matures in about 112 days |
| | Phule G 12 | 21.6 | Resistant to wilt and it matures in about 96-100 days |
| | Uday | 20 | Tolerant to root rot and wilt, stunt. it matures in about 120-130 days |
| | JG -11 | 15 | Semi spreading, bold seed, resistant to wilt and root rot, matures in about 97 days |

*Contd...*

**Contd...**

| Crops | Varieties | Yield (q/ha) | Salient Features |
|---|---|---|---|
| | JGG-322 | 17 | Medium tall plant, medium bold seed, resistant to wilt, matures in about 105 days |
| | JG 218 | 18 | Tolerant to wilt and it matures in about 115 days |
| | Vijay | 20 | Resistant to wilt and it matures in about 105-140 days |
| | ICCV 10 | 20 | Resistant to fusarium wilt and dry root rot. It matures in about 96 days |
| | ICCV 2 | 12.7 | Resistant to wilt and botrytis gray mould, matures in about 80-90 days |
| **Lentil** | K-75 | 14.0 | Semi spreading, seeds grey mottled bold (2.7g/100 seed), matures in about 135 days |
| | Narendra Massor-1 | 15-20 | Bold seeded, resistant to rust and tolerant to wilt. Maturity duration-125-130 days |
| | Pant L-406 | 20-25 | High yielding, resistant to rust and some races of wilt |
| | VLmasoor-4 | 10-12 | Suitable for hill regions, tolerant to wilt and rust, seed small and black |
| | Subrata | 12-18 | Large seed, rust tolerant and it matures in about 120-125 days |
| | Asha (B-77) | 12-17 | Rust resistant, small seed, maturity duration 120-125 days |
| | PL-06 and PL-08 | 11-15 | Preferred by the snacks industry. Maturity period 115-130 days |
| | HUL-57 | 14 | Preferred by the snacks industry an good for late conditions |
| | NDL-1 | 10-14 | Good for timely sown conditions, medium duration |
| | Moitree | 11-14 | Good for timely sown conditions, medium duration |
| **Pigeon Pea** | ICPH-8 | 24.0 | Plant indeterminate, small seed (7.7 g/100 seed), matures in about 115-135 days |
| | ICPL87 (Pragati) | 11.4 | Determinate, semi-spreading, seed light brown, medium bold (8.6g/100 seed), matures in about 116 days |
| | UPAS-120 | 11-15 | Tolerance to pod borer, early type variety |
| | T-7 | 20.0 | Suitable for intercropping with cereals, less branching |
| | Bahar | 25.0 | Resistant to sterility mosaic and suitable for rabi planting (late type) |
| **Lathyrus** | Pusa-24 | 10.65 | Old variety released by IARI |
| | Nirmal (B-1) | 14.94 | Highly adaptable to rice-utera cropping system, less BOAA content |
| | Bio L212 | 13.21 | A somaclone of Pusa-24, content very low BOAA content |
| | Racan | 15.6 | High yielding, low BOAA content |
| **French** | Him-1 | 15-20 | Maturity duration -75-80 days |
| **bean** | VL-63 | 12.0 | Light red seed with red patches, matures in about 73 days |

*Contd...*

**Contd...**

| Crops | Varieties | Yield (q/ha) | Salient Features |
|---|---|---|---|
| | PDR-14 | 18 | 125 Days red-white variegated seeds, matures in about 125 days |
| | HPR-35 | 14.0 | Medium maturity red seeds with purple streaks, matures in about 73 days |
| | | | **Oilseeds** |
| **Mustard** | M-27 | 12.0 | Maturity in 90-95 days,oil content 44.6 per cent, suitable for zero tillage operation in paddy fellow land under rainfed condition |
| | TS 29 | 12.0 | Matures in abount 85-90 days, oil content 44.0 per cent, suitable for rainfed condition |
| | Agrni (B-54) | 8-10 | Performed very well under Rainfed condition Av. Maturity 70-80 days. Oil content 41 per cent. |
| | Vinay (B-9) | 12-14 | Maturity duration -90-95 days, oil content-46 per cent |
| | Varuna | 20 | Suitable for irrigated as well as rainfed condition, maturity duration-125-130 days, Bold seeded type, oil content-39.9 per cent |
| | Kranti (PR-15) | 15-18 | Suitable for Irrigated condition, maturity-135 days, oil content-40 per cent |
| | Pusa Bold | 18 | Suitable for Irrigated condition, maturity-110-140 days, oil content-40 per cent |
| | Vardan | | |
| **Sesamum** | TKG-21(JT-21) | 8-10 | White seeded type, tolerant to Antigastra, bacterial and Cercosporaleaf spot. Maturity duration-75-78 days |
| | TKG-22(JT-22) | 6-10 | White seeded type, Tolerant to Phytophthora blight, maturity duration-76-81 days |
| **Soybean** | Pusa -24 | 30-35 | Resistant to bacterial pustules, matures in 120 days |
| | Pusa-16 | 25-30 | Purple flower, resistant to rhizoctonia and bacterial pustules, matures in 115 days |
| | JS-335 | 25-30 | Early maturity(99 days) good germination resistance to major diseases and pest |
| | MACS-124 | 25-35 | Rainfed conditions under assured rainfall areas (with 2-3 protective irrigation if required) in kharif season, matures in about 90-105 days |
| | JS 80-21 | 20 | Plant erect tall (65 cm) and purple flowered, seedling marked with purple pigmentation at basal protein, matures in about |
| | Indira soya-9 | 22.5-23.5 | Moderately resistant to stem tunneling and girdle infestation and leaf folder, matures in about 106 days |
| | Shilajeet | 20-25 | Resistant to bacterial pustules, matures in about 105 days |
| | PK-472 | 30 | Resistant to bacterial pustules, moderately resistant to YMV and matures in about 105 days |
| | VL Soya-2 | 25 | Resistant to bacterial pustule, matures in about 115 days |

*Contd...*

***Contd...***

| Crops | Varieties | Yield (q/ha) | Salient Features |
|---|---|---|---|
| | Pratap Soya 1 | 30 | Resistant to girdle beetle and moderately to stem fly and defoliators, matures in about 96-104 days |
| **Ground-nut** | ICGS-76 | 15-20 | High yielding Virginia bunch variety, matures in about 120 days |
| | ICGS -44 | 20 | A high-yielding spanish-type variety, Matures in 120 days in the post rainy season |
| | Girnar-1 | 15-18 | Early maturing multiple resistant cultivars suitable for rainfed condition |
| | ICGS-11 | 20 | Dark green leaflet; compact plant; pod 2 seeded, constriction reticulation less prominent and beak absent |
| | TG-38 | 27.6 | Tolerant to stem rot; suitable for *rabi*-summer season |
| | TG-26 | 15-20 | High yielding, earliness with fresh seed dormancy |
| | TAG-24 | 16.8 | High harvest index, early with 2-3 seeded; tolerant to BND leaf spot, jassid and caterpillar |

## Nutrient Management

Productivity of acid soils of Manipur is very low due to acute deficiency of some essential nutrients *viz.*, P, Zn, B and toxicity of Al and Mn. Inadequate and imbalanced use of fertilizers to the crops is also responsible for low system productivity and decline in soil organic carbon and fertility. In this situation, soil fertility and agro-ecosystem productivity can be maintained or even improved through the approach of integrated soil fertility management (ISFM). The principal elements to be considered while practicing ISFM are:

- ✰ Judicious use of fertilizers and soil amendments,
- ✰ Soil conservation to reduce erosion, maintain soil organic matter, and enhance water and nutrient availability to crops, and
- ✰ Cultivation of crop species, varieties in cropping systems that make optimal use of resources for food production.

Fertilizer use is inadequate, imbalanced and poorly managed in Manipur. The resource poor farmers of the state do not have money to buy fertilizers, which are often considerably very expensive because of poor transportation and infrastructure. Besides this, chemical fertilizers exclusively cannot sustain the productivity and keep soil fertility to desired level. Hence, a suitable integration of all the available nutrient sources can constitute effective strategy of soil fertility management for the low input farmers. The nutrient sources including locally available resources for sustainable crop production are:

- ✰ **Chemical fertilizers:** Besides expansion of irrigated area, use of high yielding crop varieties and other production input, chemical fertilizers have played a major role in enhancing food production. However, the farmers of hilly areas of Manipur have a mindset against the use of fertilizers and so fertilizer consumption rate is only 14189 t/year (2005-06). Sustainable growth of the agricultural production system cannot exclude fertilizer completely. In this context, increasing fertilizer use efficiency, adoption of better application techniques and reduction of nutrient losses are crucial. For correcting soil nutrient imbalances integration of chemical fertilizers along with other amendments is must.

- **Organic sources:** Lot of information on organic manure induced improvement of soil fertility and sustainable crop production is available. Integrated nutrient supply including the use of chemical fertilizers with organic sources like green manure, FYM, crop residue, bio-fertilizers etc. helps not only in bridging the gap between nutrient removal and addition, but also in enhancing nutrient response, use efficiency and in improving crop productivity and quality as well. Long-term fertilizer experiments in different agro-ecological zones of India clearly showed that even recommended doses of NPK fails to sustain crop productivity and soil quality. However, combined use of fertilizers and FYM could produce higher crop yield beside improvement of soil fertility.
- **Bio-fertilizers:** These are the cultures of microorganisms used for inoculation of seed or soil or both to mobilize nutrients from unavailable to available form through biological processes. There are mainly three types of bio-fertilizers:

  I. Nitrogen fixing bio-fertilizers (NBF),

  II. Phosphate solubilizing bio-fertilizers (PSB),

  III. Potash solubilizing biofertilizers (KSB).

  The NBF fix nitrogen into soil by reducing atmospheric nitrogen. Globally, biological N fixation potential is reported to be 140-170 million tonne N/year; whereas N input through fertilizer is only 70 million tonne N/year chemically. The types of NBF commonly available in form of:

  *a)* *Rhizobium* which is symbiotic N-fixer used to inoculate specific leguminous crop. This type of NBF is most popular.

  *b)* *Azospirillum* which is non crop specific and is used mainly for cereal crops, and

  *c)* *Azotobacter* which is non-symbiotic N fixer used for inoculating wide range of crops.

  The PSB contains live cells of phosphate solubilizing bacteria (*Pseudomonas* and *Bacillus*) and fungi (*Penicillium* and *Aspergillus*). The performance of PSB is not consistent as these cultures are likely to be more useful in soils with high organic matter content and favourable environment for those microorganisms. The KSB generally contains *Frateuria aurentia* whih is very useful for potash exhaustive crops.
- **Green manuring:** It involves the soil incorporation of undecomposed green plant tissues for the purpose of improving soil fertility. Broadly, two types of green manuring are there:

i) ***In situ* green manuring** - green manure crops are grown either as pure crop or inter crop and buried in the same field which is to be green manured. The most common crops for in *situ* green manuring are sunhemp (*Crotolaria juncea*) and dhanicha (*Sesbania aculeata*).

ii) **Green leaf manuring** - refers to turning the soil green leaves and tender twigs collected from shrubs and trees grown on bunds, wasteland and forest area etc. The shrubs and trees used are Gyricidia (*Glyricidia maculate*) and Karanj (*Pongamia pinnata*).

- **Crop residue:** Crop residues are the plant parts or crop remains left after harvesting and/or processing of the produce. Annually, ~ 355 million tonne of crop residues are available in India. Out of this 180 million tonne can be utilized to supply 3.5 million tonnes of plant nutrients. Rice, wheat and sugarcane are potential crops which ensure availability of large amount of crop residues. These residues can be used as animal feed, mulch to other crops or can be recycled as compost. Burning of crop residues is another practice to quickly dispose the crop residues. However, it is wasteful practice as there is loss of nutrients beside environmental pollution.
- **Composting:** Composting is the best method of recycling agricultural waste. Enrichment of the nutrient values of compost can be done with mixing rock phosphate and microbial inoculants. Now-a-days earthworms are being used for quick decomposition of farm waste for composting. The product is popularly known as vermicompost. Worm species commonly used for composting are: *Eisenia foetida, Eudrilus engeniae* and *Perionyx excavatus*. *Eisenia foetida* is reported to be the most efficient one.
- **Urban waste:** City and urban waste can be used as source of plant nutrients after proper treatment. Urban sludge which contains various nutrient elements including micronutrients can improve soil physical conditions as well.

## Soil Management

### Soil Acidity Reclamation

Soil acidity is a major contributing factor that affects crop productivity in Manipur. Soil acidity can also be ameliorated with minimally processed rock phosphate, basic slag and paper mill sludge. The use of rock phosphate and basic slag containing considerable amount of P would especially be promising, since low P availability is a primary constraint in acidic soils. Large deposits of limestone (46 million tonne) are also there in Manipur that can be exploited for agricultural production through minimum processing at affordable cost. The available technologies that could be adopted to achieve desirable success in liming acid soils are:

- Source of lime - rock phosphate, basic slag, paper mill sludge, limestone.
- Lime application rate - 2-5 q/ha (5-10 per cent of lime requirement as determined in soil testing laboratories).

- ✰ Method of lime application - be applied in furrows along with the basal dose of fertilizers manually or with the help of fertilizer drill. For any cropping system lime is to be applied to every crop.
- ✰ Choices of crops - lime responsive crops are rapeseed, green gram, soybean, groundnut, lentil, pea, linseed, pigeon pea, maize etc. Paddy being the most commonly grown crop in the region did not come up well with liming due to self-liming effect under flooded condition.

The second element of ISFM is soil management to conserve and enhance soil fertility, including erosion control, maintenance of soil organic matter, and enhancing nutrient bioavailability by promoting beneficial root symbiosis via biological N fixation and mycorrhizal associations. Although the critical importance of these management tools has been well known, their use is limited among the farmers. Because of ignorance and lack of awareness for soil resources many poor farmers use them abusively, without considering the long-term consequences of practices such as deforestation, residue burning, and cultivation of steep slopes without runoff barriers. Soil resources may be devalued because of the transitory nature of the cropping system, such as in the 'slash and burn' agriculture particularly in the hill districts or because the land is not owned by the farmer.

The third element of ISFM is the cultivation of crop species, varieties in cropping systems that make optimal use of soil resources for food production while conserving soil fertility. Although crop species and cropping systems vary substantially in their ability to produce food in marginal soils, the adoption of new crop species and cropping systems is subject to some of the same socio-economic barriers along with the cultural attachments to specific foods. Most of the farmers grow such crops which are used for their daily consumption. Strong preferences may exist for specific grain types (*e.g.*, sticky rice favoured in Manipur) within a crop species, regardless of yield advantages to introduced types. The cultivation of more nutrient efficient crop species will probably be an important element of the adaptation. However, these changes will likely be slow for traditional agricultural communities. The prospect of cultivating new genotypes of existing crops with greater productivity in infertile soils is considerably more promising, especially if such genotypes have agronomic and grain characteristics that are similar to traditional landraces. Substantial genotypic variation exists in crop species for tolerance to Al toxicity, Mn toxicity and low P, and this variation has been deployed through crop breeding programs. Selection for root traits that increase the acquisition of limiting nutrients such as P in crop plants will be important objective of breeding programs for Manipur condition.

## Water Management

### *In situ* Water Management

#### (a) Agronomic Measures

This measures adopted when the slope is gentle *i.e.* less than two percent.

- ✰ **Contour cultivation:** Ploughing and sowing of crops across the slope reduces soil erosion and act as an obstruction to runoff, providing adequate time for water to enter into the soil and reduce soil loss.

☆ **Mulching**: Mulching with plant materials reduces soil loss up to 43 times compared to bare soil and 17 times compared to cropped soil without mulches. Various types mulches that can be used for rain water conservation are as follows:

(i) **Dust mulch:** a fine loose dry layer of surface soil maintained by cultivation under the assumption that it will prevent evaporation of soil moisture and increase water retention.

(ii) **Stubble mulch:** Stubble mulching is leaving the plants stubble from the certain height. It allows for less erosion, enhance water holding capacity and organic carbon pool. eg. Wheat, Paddy, Maize stubble etc.

(iii) **Plastic mulch:** Plastic mulch is a product used, in a similar fashion to mulch, to suppress weeds and conserve water in crop production.

**Plastic (Textile) Mulch in Broadbean at ICAR Langol Farm.**

**Plastic (Textile) Mulch in Banana at ICAR Langol Farm.**

(iv) **Vertical mulching:** Sub soiling for breaking of hard pan by heavy plough at 40-60 cm depth of soil. It increases the root penetration, aeration, and water percolation. This is mostly practiced in coffee garden.

(v) **Straw mulch and crop residues:** Application of crop straw and residues on the surface of soil resulting in storing more rain water in soil by reducing runoff, erodibility, soil compaction and evaporation rate from soil. It increases infiltration rate, microbial biomass, soil aggregation, nutrient retention and aeration in the rhizosphere, resulting in direct and indirect increase in agronomic productivity, water use efficiency (WUE), nutrient use efficiency (NUE) and energy use efficiency (EUE).

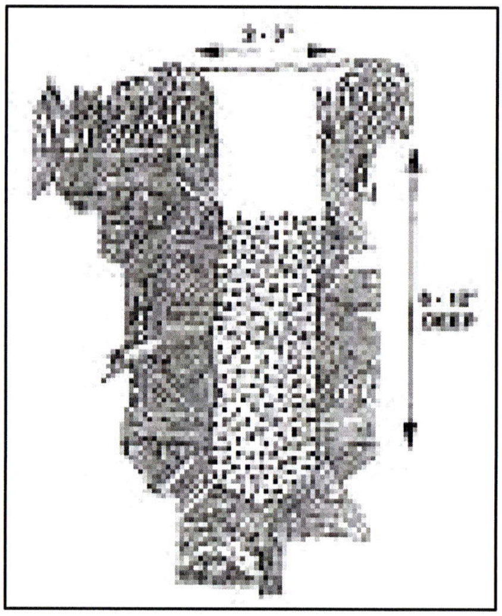

**Vertical Mulching.**

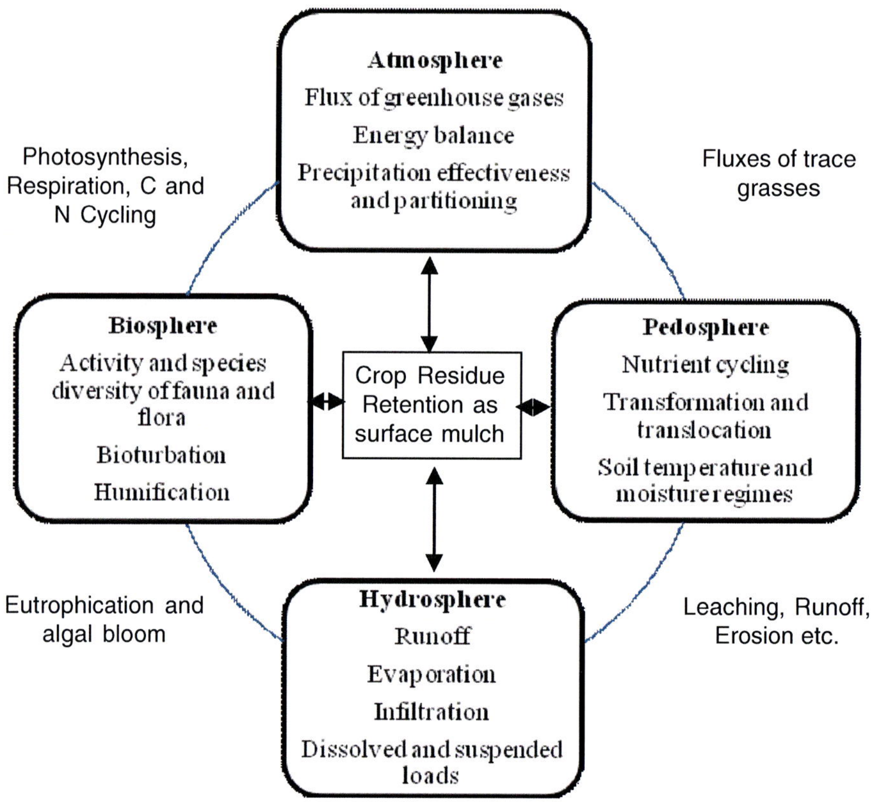


**Impact of Soil Organic Carbon on Atmosphere, Pedosphere, Hydrosphere and Biosphere.**

- **Check dams:** Check dams are constructed across gullies to reduce the velocity of runoff, heal the gully, store water for use by livestock and recharge groundwater in wells lower down.
- **Farm ponds:** Farm ponds of normal size 100 – 300 $m^3$ may be dug to store 30 per cent of runoff. Seepage loss can be controlled by use of bentonite which has good sealing properties.

**Farm Ponds**

- **Percolation ponds:** Flowing rivulets or big gullies are obstructed and water is ponded. Water from the ponds percolates into the soil and raises the water table of the region. The improved water level in the wells lower down the percolation tanks are used for supplemental irrigation.
- **Broad bed and furrow method (BBF):** The seed bed is made into wide beds of 1.05 m alternated with furrows of width 0.45. This method is quite suitable for medium to deep black soils (Vertisols) and rainfall 700 –1300 mm with Maximum up to 5 per cent Slope.

**Groundnut Crop Sown under BBF.**

- ☆ **Inter row and plot water harvesting:** In areas of high rainfall, there is possibility of occasional waterlogging and yield of crop like maize is affected. Growing maize on beds and rice in furrows helps in increasing the yields of both the crops. The excess water collected on beds is stored in furrows which is beneficial for rice.

## (b) Mechanical Measures

Mechanical measures are adopted to supplement the agronomical measures when later alone are not adequate.

**Graded bunding:** Graded bunding is suitable in situations where the rain water is not readily absorbed either due to high rainfall or low intake of the soil, graded bunding is recommended.

- ☆ **Broad base and terrace method:** A terrace is a combination of ridge and channel built across the slope on a controlled grade. These terraces have wide base and low height of ridge and usually formed with machinery. It is useful for deep black soils with high clay content develop deep cracks in summer.
- ☆ **Bench terracing:** On steep slope and undulated land, intensive farming is possible only with bench terracing. It is usually practiced on slopes 16 – 33 per cent. The field is made a series of level strips or platforms across the slope of the land.
- ☆ **Zing terracing:** Zing terracing is adopted in lands with 3-10 per cent slopes and bench terracing is recommended on steeper slopes. It construct medium to deep soils in moderate rainfall areas.

## (c) Agrostological Measures

- ☆ **Vegetative barrier - *Vettiveria zelanica* (khus grass):** Vetiver hedges have been reported to conserve 21.80 per cent, 20.98 per cent, and 19.77 per cent soil moisture at various stages (30, 60 and 90 DAS) of growth and produce higher yield of maize. In dryland farming, vetiver - hedging for moisture conservation and higher yield of the intercrops has been recommended.

**Runoff Water Management.**

- ☆ **Watershed**: Watershed is an area of land where rainfall is collected and drains through a common point. Watershed management is the integration of technologies with in the natural boundaries of a drainage area for optimum development of land, water and plant resources to meet the basic needs of the people and animals in a sustained manner.

## Disease Management of some Important Crops

### A. Rice Blast

Blast disease caused by the fungus *Magnaporthe grisea* Cav.[anamorph *Pyricularia grisea* (Cooke) Sacc.]. The typical symptoms of blast appear on the leaves, but the leaf sheath, rachis, nodes, and even glumes are infected. On the leaves, the symptoms first appear as small bluish flecks, about 1-3 mm in diameter. The lesions enlarge to several centimeters long and are typically spindle- or diamond-shaped, with gray or almost straw colored centre and brown margin. Lesions expand rapidly and tend to coalesce, leading to complete drying of infected leaves. The most characteristic symptoms appear on the culm. Infection to the neck node produces triangular purplish lesions, followed by lesion elongation to both sides of the neck node- symptoms which are serious for grain development. When young neck nodes are invaded, the panicle becomes white in color. Later infection causes incomplete grain filling, and poor grain quality or chaffy panicles.

**Control Measures**

- ☆ Avoid excess nitrogen and skip final nitrogen application in blast infected field.
- ☆ Treat seed with Tricyclazole 75WP or Carbendazim 50WP @ 2g/Kg seed.
- ☆ Foliar spray can be done with Tricyclazole 75 WP @ 0.6 g or Isoprothiolane 40EC @ 1.5ml or Kasugamycin 40EC @ 2ml or Ediphenphos 50EC @ 1ml or Iprobenphos 48EC @ 2ml or Propiconazole 25 EC @ 1ml or Carbendazim 50WP @ 1 g or Thiophanate methyl @ 1g per litre of water.

### B. Sheath Blight of Rice

Sheath blight is a fungal disease caused by *Rhizoctonia solani*.Symptoms of the disease usually appear when plants are in the late-tillering or early internode elongation growth stages. Small, water-soaked spots (about 1 cm long) first appear on the leaf sheath at or above water level as water soaked, circular to oblong, ellipsoid to ovoid shapes and somewhat irregular greenish grey in colour, Older lesions are elliptical or ovoid with a grayish white center and light brown to dark brown margin They enlarge to approximately 1 cm in width and 2-3 cm in length. Lesions may coalesce forming bigger lesions with irregular outline and this lesion interrupts the flow of water and nutrients to the leaf tip and the tip may die. As the plant grows and the canopy closes, the humidity inside the canopy increases. In this humid environment, the fungus grows upward inside the plant and on the plant's surface, causing new lesions. The fungus also can spread to nearby plants. Severely damaged plants may lodge. Eventually the hole sheath rots and the affected leaf can easily be pulled off, from the plant. In severe cases all the leaves of a plant are blighted resulting in death of the plant. Severely infected plants produced poorly filled or empty grains, especially those on the lower portion of the panicles. Poorly developed grain usually breaks up during milling, thus reducing quality of rice.

**Control Measures**

- ☆ Avoid high seeding rates in nursery and denser planting of seedlings in field of rice, cause less air movement among the plants. This results in more moisture on the plants from dew and rain and promotes conditions that favor sheath blight development. Transplanting rice seedlings at spacing of 25 x 25 cm. against 15 x 15 cm. reduces favorable effect of high N application in disease development could be reduced.
- ☆ Keep the field clean by destroying grasses and other collateral hosts to reduce the inoculums.
- ☆ Split nitrogen applications and avoid excessive nitrogen rates. Excessive nitrogen promotes succulent, dense growth of rice that encourages sheath blight.

## C. Bacterial Leaf Blight of Rice

Bacterial leaf blight is caused by *Xanthomonas oryzae pv. oryzae*. The disease is observed on both seedlings and older plants. On seedlings, infected leaves turn grayish green and roll up. As disease progresses, leaves turn yellow to straw-colored and wilt, leading whole seedlings to dry up and die. This expression of the disease is known as "kresek". On older plants, lesions usually develop as water-soaked to yellow-orange stripes on leaf blades or leaf tips or on mechanically injured parts of leaves. Lesions have a wavy margin and progress toward the leaf base. On young lesions, bacterial ooze resembling a milky dew drop can be observed early in the morning. The bacterial ooze later on dries up and becomes small yellowish beads underneath the leaf. Old lesions turn yellow to grayish white with black dots due to the growth of various saprophytic fungi. On severely infected leaves, lesions may extend to the leaf sheath.

**Control Measures**

- ☆ The use of balanced amounts of plant nutrients, especially nitrogen.
- ☆ Ensuring good drainage of fields (in conventionally flooded crops) and nurseries.
- ☆ Keep fields clean - removing weed hosts and plowing under rice stubble, straw, rice rations and volunteer seedlings, which may constitute sources of inoculum.
- ☆ Allowing fallow fields to dry to suppress inoculum in the soil and plant residues.
- ☆ Up to 95 per cent of the seed infection can be eradicated by soaking the seeds for 12 hours in 0.025 per cent solution of Agrimycin-100 (15 per cent streptomycin +1.5 per cent terramycin) plus 0.05 per cent wettable ceresan and then transferring the seed to hot water at 52-53 °C for 30 minutes.

Overnight soaking of infected seeds in 100ppm streptocycline (streptomycin 12 per cent + chlorotetracycline hydrochloride 1.5 per cent) solution can also effectively eradicate the seed infection.

- Spraying twice with 250ppm of Agrimycin 100 can effectively reduce the disease intensity and has been advocated for checking secondary spread of the disease.
- Application of Klorocin (stable bleaching powdcry @12.5 kg/ha at nursery and 10 days after transplanting can reduce the disease.

## D. Turcicum Leaf Blight of Maize

Turcicum blight is caused by *Helminthosporium turcicum* (Teleomorph: *Setosphaeria turcica*). Symptom appear as slightly oval, water-soaked, small spots produced on the leaves. These grow into elongated, spindle-shaped necrotic lesions. They may appear first on lower leaves and increase in number as the plant develops, and can lead to complete burning of the foliage. Turcicum leaf blight (or northern leaf blight) occurs particularly in areas where high humidity and moderate temperatures prevail during the growing season. When infection occurs prior to and at silking conditions are optimum, it may cause significant economic damage.

## E. Southern Maize Leaf Blight or Maydis Leaf Blight

Turcicum blight is caused by *Bipolaris maydis* (Teleomorph: *Cochliobolus heterostrophus*). Young lesions are small and diamond shaped. As they mature, they elongate. Growth is limited by adjacent veins, so final lesion shape is rectangular and 2 to 3 cm long. Lesions may coalesce, producing a complete burning of large areas of the leaves. Maydis leaf blight (or) is prevalent in hot, humid, maize-growing areas. The fungus requires slightly higher temperatures for infection than *E. turcicum*; however, both species are often found on the same plant.

### Control Measures

- Infected crop debris should be incorporated in soil by deep ploughing.
- Spray Dithane M-45 or Zineb @ 2.4-4.0 g/liter (2-4 applications) at 8-10 days interval after first appearance of disease symptoms.

## F. Banded Leaf and Sheath Blight

Banded leaf and sheath blight is caused by the fungus *Rhizoctonia solani* f. sp. *sasakii*. The disease develops on leaves and sheaths. Symptoms are characteristic concentric spots that cover large areas of infected leaves and husks. The main damage in the humid tropics is a brownish rotting of ears, which show conspicuous, light brown, cottony mycelium with small, round, black sclerotia.

### Control Measures

- Seed treatment with peat based formulation (*Pseudomonas flourescence*) @ 16 g/kg of seed or as soil application@ 7g/liter of water (soil drenching) or foliar spray of sheethmar (validamycin) @ 2.7 ml/liter water.

## G. Early and Late Leaf Spots of Groundnut

### Early Leaf Spot

It is caused by *Cercospora arachidicola*. It develops small necrotic flecks, that usually have light to dark brown centers, and a yellow halo. The spots may range from 1 mm - 10 mm in diameter and are circular to irregular in shape. Sporulation is on the adaxial (upper) surface of leaflets.The color of spots on the lower leaf surface is brown.

### Late Leaf Spot

It is caused by *Phaeoisariopsis personata*. Its earlier name was *Cercosporidium personatum*. It develops small necrotic flecks that enlarge and become light to dark brown. The spots are usually circular in shape The yellow halo is either absent or less conspicuous in late leaf spot. Sporulation is common on the abaxial(lower) surface of leaves. The color of spots on the lower leaf surface is black.

### Control Measures

- ✰ A crop rotation of cereal - cereal - groundnut and burying all groundnut crop residues by deep plowing will reduce initial inoculum.
- ✰ Adjust the date of sowing to avoid conditions favourable for disease development
- ✰ Multiple applications of a fungicide such as benomyl, captafol, chlorothalonil, copper hydroxide, mancozeb, or sulphur fungicides may control early and late leaf spot. However, carbendazim (0.05 per cent) controls both leaf spots very effectively. Three sprays of 0.2 per cent chlorothalonil at intervals of 10 - 15 days starting 40 days after germination up to 90 days provides effective control to early and late leaf spots and rust.
- ✰ Grow cultivars tolerant to late leaf spot: ICGV 87160 or ICGV 86590.

## H. Groundnut Rust

Rust (*Puccinia arachidis*) is identified by the appearance of orange pustules (uredinia) on the abaxial (lower) surface of leaves and reddish -brown urediniospores (uredospores). Symptoms are mainly confined to leaflets but pustules can be seen on all the aerial parts of a plant except the flower. Brown to dark reddish brown pustules appear on the lower surface with the upper surface developing yellow, chlorotic spots with necrotic areas in the center. At a late stage, the primary pustules are surrounded by secondary sori.The uredinias are usually circular, 0.3 mm – 1.0 mm in diameter, and can develop on all the aerial parts of the plant except flowers and pegs.

### Control Measures

- ✰ A cereal - cereal - groundnut crop rotation and removal of volunteer groundnut plants from the field will help to check rust inoculums build up.
- ✰ Adjusting the sowing time to avoid the most conducive environmental condition for rust development (*i.e.*, high humidity, cloudy weather) to help reduce damage caused by rust.

- ☆ Sprays of Bordeaux mixture and dithiocarbamate have been found effective to control rust and late leaf spots. Chlorothalonil 0.2 per cent spray has been found effective against rust and late leaf spot, when sprayed 30 days after germination till 15 days before harvesting at regular 10 - 15 day intervals. Calixin is effective against rust but not against leaf spots, whereas benomyl is effective against leaf spots but not against rust.
- ☆ Grow resistant cultivars: ICGV 87160 or ICGV 86590.

## I. Powdery Mildew of Pea

Powdery mildew is caused by *Erysiphe pisi* fungus which overwinters on infected plant debris and in alternate hosts. First symptoms appear on the upper surface of the foliage as very small, slightly discolored spots. These spots give rise to white powdery areas which continue to enlarge. As they do so, the white powdery color and powdery appearance become more pronounced. Tissue beneath these areas may turn purplish, then brown. Multiple infections may cover the entire above ground plant. Severely infected plants are unthrifty and have poor yield and quality. Small, oval, black fruiting structures may form in mature lesions.

### Control Measures

- ☆ Cleistothecia surviving in soil may serve as an important source of primary inoculum, therefore, field sanitation is important to check infection from soil borne inoculum. The debris of diseased crop should be burnt in the field.
- ☆ Spray the crop on the appearance of disease with Dinocap (50 ml/100 L water) or Sulfex (200 g/100L water) or Carbendazim 50 WP (50 g/100 L water) or Bitertanol (50 g/100L water) at 10-15 days interval depending on the severity of disease.

## J. Asochyta Leaf and Pod Spot of Pea

Three fungi *Mycosphaerella pinodes, Phoma medicaginis* var. *pinodella*, and *Ascochyta pisi* incite Ascochyta diseases of pea. Typical lesions on foliage and pods are larger (5-8 mm) in diameter and are sunken and tan, with a dark distinct border. Lesions are oval on foliage and pods, elongated on stems. The centers of lesions caused by all three pathogens show dark pimple like structures containing the early dispersed spores which spread the disease. The pathogens are carried in infected seed and overwinter in infected plant debris.

### Control Measures

- ☆ Use disease free seed.
- ☆ Where possible, pea crops should not be grown more than once every three or four years in the same field. This will reduce the risk of infection from pea residue or soil-borne spores in that field.
- ☆ Plough pea refuse deeply and completely.

## K. Fusarium Root Rot of Pea

The disease is caused by *Fusarium oxysporum* f.sp. *pisi*. Initial symptoms often appear near the area of seed piece attachment and consist of slender, light brown lesions along the taproot and on the side roots. These lesions enlarge and run together until the main roots become completely covered, shrunken, and dark brown or light red in color. This discoloration and shrinking continues upward to 1-2 inches above the soil line. The shrinking is caused by the collapsing of dead cortical cells. The vascular tissue also shows red discoloration extending upward one to three nodes above the soil line. Diseased plants appear unthrifty, variously dwarfed depending upon the severity of infection, and may wilt and die.

**Control Measures**

- ✰ Rotate crops.
- ✰ Plant infested fields early or use an early maturing cultivar.

## L. Bacterial Blight of Pea

Bacterial blight is caused by *Pseudomonas syringae* pv. *phaseoli*. Lesions appear on all aboveground parts. They begin as small, water-soaked, oval spots. Individual leaf and stipule lesions enlarge to approximately 3mm in diameter with an angular shape. Pod lesions are 6mm or more in width, and stem lesions lengthen to resemble stem streak. Lesions vary in color, but on foliage they generally turn brown, shiny, and translucent. Pod lesions, which may extend along either suture, slowly change from water soaked to greasy appearance with purplish brown margins. Peas inside infected pods may develop brown discoloration. Long stem lesions turn brown with age. The bacterium overwinters in infected plant debris and in the seed coat.

**Control Measures**

- ✰ Use disease free seed.
- ✰ Follow atleast three years crop rotation with non-host crops.

## M. Aphanomyces Root Rot of Pea

The disease is caused by *Aphanomyces euteiches* f.sp. *pisi* fungus. First signs of infection are long, soft, water-soaked areas on the surface of the lower stem and root. These areas soon become light tan and spread to cover the whole root system. Because the disease kills branch roots, only the water and food conducting tissues of the upper taproot remains intact when the plant is pulled. Diseased plants are stunted and weakened and in severe cases they wilt, turn yellow, shrivel and die prematurely.

**Control Measures**

- ✰ To avoid soil infestation, use long rotations.
- ✰ Select field with well drained friable soils.

## N. Alternaria Leaf Spot of Mustard

Leaf spot of mustard caused by Alternaria brassicae appear as small, minute dark colored area, which expand very fast to form circular lesions upto 1-2 cm diameter.

In moist weather the fungus may appear as a bluish growth in the center of the spot. Concentric rings formed in the lesions. In temperate countries, this fungus is seed borne and causes shriveling of seed and low germination. Spores and mycelium in diseased plant debris also serve as mean of spread of disease

**Control Measures**

- ☆ Leaf spot can be effectively controlled by regular spray of fungicides. 2-4 sprays of 0.2 per cent Mancozeb (Dithane M-45),Dithane Z-78, Blitox-50 at 15 days interval starting 45 days after sowing. Ridomil MZ is highly effective to control the leaf spot of mustard.
- ☆ Plant debris should be collected and burnt after the harvest of the crop.

## O. White Rust of Mustard

The disease causes local and systemic infection in the plant. All parts of plant except roots may show symptoms. In local infection, isolated pustules or sori develop on leaves and stems. These pustules or blisters are raised, shiny white areas, measuring 1-2 mm in diameter. Host epidermis ruptures earlier or sometimes after the pustules are fully formed. Blisters are also formed on the floral parts. Due to hypertrophy and hyperplasia in the tissue the floral parts show swelling and distortion in systemic infection. Fungus perpetuates through oospores in soil in diseased plant debris diseased species. Perennial weed host plant serve as a primary source of inoculum and secondary spread is carried out by sporangia and zoospores

**Control Measures**

- ☆ Follow clean cultivation including certified and healthy seed.
- ☆ Crop rotation should be used to control the soil born primary inoculum. The disease can be control by spray of 0.8 per cent Bordeaux mixture, 0.3 per cent Difolaton, and 0.1 per cent Ridomil at 8-10 days interval.

# Insect-Pest Management of some Important Crops

## Rice

Major insect pest of rice in Manipur are the stem borer, leaf folder, gundhibug, gallmidge and whorl maggot

### A. Rice Stems Borer *Scirpophaga incertulas*

***Symptom of Damage***

- ☆ Presence of brown coloured egg mass near leaf tip.
- ☆ Caterpillar bore into central shoot of paddy seedling and tiller
- ☆ Causes drying of the central shoot known as "dead heart" Grown up plant whole panicle becomes dried "white ear

***Management***

- ☆ Since the eggs are laid near the tip of the leaf blade, clipping the seedlings before transplanting reduces the carry-over of eggs from seed bed to the transplanted field.

- ☆ Harvesting at ground level or ploughing after harvesting remove majority of larvae and pupae.
- ☆ The use of light traps was recommended earlier but now this method is not advocated as many beneficial insects are also attracted and killed.
- ☆ **Use of Pheromone traps:** sex pheromones are utilized for mass trapping of male moths.
- ☆ Application of carbofuran 3G @ 15 kg/ha at the time of transplanting followed by six releases of *Trichogramma japonicum* (Local) @ 50,000/ha at fortnightly intervals from 45 days after transplanting
- ☆ **ETL recommended**– 5 per cent dead hearts or 1 per cent white ear or one egg mass or one adult/Sq.mt.

### B. Leaf Folder *Cnapholocrocis medinalis*

#### *Symptom of Damage*

The larva rolls the leaf blade by fastening its edges and sometimes even fastening the leaf tip to the basal part of the leaf blade and feeds from inside by scraping. In a severely infested field the whole crop gives a sickly appearance with white patches. The infestation at boot leaf stage of the crop sometimes results in heavy loss of grain yield.

#### *Management*

- ☆ **Nursery** - Carbofuran 3G @ 42 Kg/ha or Phorate 10 G @ 12.5Kg/ha to be applied 7 days after germination of seed or spraying of Monocrotophos 1.6 ml or Chlorpyriphos 2.5 ml/li or Quinalphos 2ml/li at 10 days interval starting from 7 th day after germination.
- ☆ **Main Field**
  - **Transplanting to panicle initiation** - Carbofuran 3 G @ 25 Kg/ha OR Spraying of Monocrotophos 1.6 ml, or Chlorpyriphos 2.5 ml OR Quinalphos 2.5 ml or Carbaryl 3 gms or Phosphomidon 1 ml/lit water.
  - **Panicle initiation to booting** - Carbofuran 3 g @ 25 Kg/ha OR Spraying of Monocrotophos 1.6 ml or Chlorpyriphos 2.5 ml OR Quinalphos 2.5 ml or carbaryl 3 gms or Phosphomidon 1 ml/lit. of Water.
  - **Post flowering** - Spray Monocrotophos 1.6 ml or carbaryl 3 gms. OR Phosphomidon 1 ml or Cartap hydrochloride at 2 ml/lit of water to be sprayed OR Cartap hydrochloride 4 G granules @ 20 kg/ha to be applied.

### C. Gall Midge, *Orseolia oryzae*

Maggot feeds at the base of the growing shoot causing formation of a tube like gall that is similar to "onion leaf" or "Silver-shoot". Infested tillers produce no panicles.

#### *Management*

(i) Seed treatment with chlorpyriphos 0.2 per cent emulsion for 3 hours or seed mixing with either chlorpyriphos (0.75 kg a.i./100 kg seeds) or imidacloprid (0.5 kg a.i./100 kg seeds) provide protection for 30 days in the nursery.

(ii) Seedling root dip in 0.02 per cent chlorpyriphos emulsion before transplanting for 12 -14 hours gives protection for 30 days.

(iii) Removal and destruction of weeds that serve as alternate host plants.

## D. Gundhi Bug

- Sucking the sap from individual grains, which are in milky stage.
- Individual grains become chaffy
- Black spots on the grains at the site of feeding puncture.
- Buggy odour in rice field during milky stage

### *Management*

- Removal of weeds in the vicinity of paddy crop as the pest breeds on a variety of grasses prior to its migration to rice crop.
- Collection of the bugs by hand netting is suggested. Sweeping of rice plants with winnows smeared with sticky material like castor oil.
- Spray NSKE 5 per cent or Need based spraying of chlorpyriphos 20 EC 1250 ml/ha or phosalone or carbaryl or monocrotophos or cartap hydrochloride or quinalphos or fenthion at 0.5 kg a.i./ha or spray of fipronil 5 SC at 1 litre/ha.

## E. Whorl Maggot, *Hydrellia sasakii*

### *Symptom of Damage*

- Maggot feeds on the tender tissue inside the whorl
- Yellowish white longitudinal marginal blotching with hole
- Leaves shrivelled plant stunted and maturity delayed.
- Drooping of young leaves near the tip

### *Management*

- Remove the alternate hosts and adjust the time of planting (early)
- Use early maturing varieties
- Optimum recommendation of potash fertilizer
- Spraying of endosulfan or quinalphos or fenthion at 0.5 kg a.i./ha. Alternatively, application of granules of carbofuran or fenthion at 0.75 kg a.i./ha

# Maize

## A. Cob Borer *Stenachroia elongella*

### *Symptoms of Damage*

Newly emerged larvae puncture the cobs and the excrement on the cob is the symptom of damage

Some time larvae bores into the stem causing multiple damage.

***Management***

- ☆ Deep ploughing exposes the the larval and pupal stages of pest to natural predators like birds.
- ☆ Burning of crop residues or allow farm animals to graze on it kills the over wintering stages *i.e.* larvae and pupae.
- ☆ Crop should be planted early to overcome cob borer attack
- ☆ Spraying of Neemarc 3ml/lit at silking stage reduce cob borer population
- ☆ Spraying of monocrotophos (0.005 per cent) or dipel @ 1 l/ha at 80-90 DAS gives effective control

**B. Stem Borer *Chilo partellus***

***Symptoms of Damage***

- ☆ Central shoot withers and leading to "dead heart".
- ☆ Larvae mines the midrib enter the stem and feeds on the internal tissues.
- ☆ Bore holes visible on the stem near the nodes.
- ☆ Young larva crawls and feeds on tender folded leaves causing typical "shot hole" symptom.
- ☆ Affected parts of stem may show internally tunnelling caterpillars

***Management***

- ☆ Set up light trap till midnight to attract and kill the stem borer moths.
- ☆ Collect the stubbles after harvest and burn to destroy diapausing borers.
- ☆ Two spray of cypermethrin at 100ml/ha or deltamethrin at 200ml/ha or carbaryl 50 WP @1 kg (500 lit. spray fluid/ha gives effective control
- ☆ Application of carbofuran 3G @ 7.5 kg/ha is most effective to control stem borer or carbofuran 3G@ 5-6 granules per whorl at 25-30 days after germination effectively checks the stem borer.

**C. Corn Worm *Helicoverpa armigera***

Larva feeds on silk and developing grains.

***Management***

- ☆ Deep summer ploughing and destruction of pupae.
- ☆ Intercrop cotton with cowpea, groundnut, greengram and soybean to buildup natural enemies like ladybird beetles, chrysopa and syrphids etc.
- ☆ Monitoring adult insects with pheromone traps @ 4 traps per acre.
- ☆ Clipping the plant tips at 45 days after sowing to remove the eggs of larvae.
- ☆ Spraying 5 per cent Neem seed kernel suspension as repellent
- ☆ Spraying of fenvalerate 20EC @ 100ml/ha or cypermethrin 10EC @ 100ml/ha or deltamethrin 2.8EC @ 200ml/ha

**D. Aphid *Rhopalosiphum maidis***

***Symptoms of Damage***

- ☆ Yellowing of leaves
- ☆ Colonies of aphids found in central leaf whorl.
- ☆ Nymphs and adults suck the sap from leaves and young shoots.
- ☆ Exude honeydew on which a sooty mould grows giving the leaves a black appearance

***Management***

- ☆ One spray of systemic insecticides like dimethioate or monocrotophos at 0.1 per cent or with rogor 30EC @ 1 ml/lt water will controls the aphids.

**E. Pink Stem Borer *Sesamia inferens***

***Symptoms of Damage***

- ☆ Pink larva enters into the stem causing dead heart symptom.

**Management**

- ☆ Spray carbaryl 50 WP @1 kg (500 lit. spray fluid/ha or
- ☆ Spraying of monocrotophos (0.005 per cent) or dipel @ 1 l/ha at 80-90 DAS gives effective control
- ☆ Application of carbofuran 3G @ 7.5 kg/ha is most effective to control stem borer

## Groundnut

**A. Bihar Hairy Caterpillar *Spilosoma (Diacrisia) oblique***

***Symptoms of Damage***

- ☆ Young larvae feed gregariously mostly on the under surface of the leaves.
- ☆ Feed on leaves and cause loss by way of defoliation.
- ☆ In severe cases only stems are left behind.

***Management***

- ☆ Pre-monsoon deep ploughing (two/three times) will expose the hibernating pupae to sunlight and predatory birds.
- ☆ Removal and destruction of alternate wild hosts and weeds which harbour the hairy caterpillars.
- ☆ Irrigate once to avoid prolonged mid season drought to prevent pre-harvest infestation.
- ☆ Mass collection and destruction of eggs and just emerged caterpillars.
- ☆ Conserve the bio control population of spiders, long horned grasshoppers, preying mantids, robber fly, ants, green lace wing, damsel flies/dragon flies, flower bugs, shield bugs, lady bird beetles, ground beetle, predatory cricket, braconids, trichogrammatids, NPV, green muscardine fungus

- ☆ Use of NPV (nuclear polyhedrosis virus) on cloudy days at 500 LE/ha will be effective. Spraying of Bacillus thuringiensis is also advocated at 1 kg/ha where mulberry is not grown.
- ☆ Apply safe chemical insecticides at recommended doses only if the insect population crosses the ETL.
- ☆ Spraying of quinalphos 25 EC (2 ml/lit) or chlorpyriphos 20 EC (2.5 ml/lit) recommended when the caterpillars are younger.

### B. Gram Pod Borer *Helicoverpa armigera*

#### *Symptoms of Damage*

- ☆ Larvae feed on the foliage, prefers flowers and buds.
- ☆ When tender leaf buds are eaten symmetrical holes or cuttings can be seen upon unfolding of leaflets.

#### *Management*

- ☆ Deep summer ploughing.
- ☆ Intercrop one rows of red gram for every 5 or 6 rows
- ☆ Install pheromone trap @ 5/ha
- ☆ Use *Trichogramma chilonis* @ 1 lakh/ha or *Chrysoperla carnea* @ 50000/ha at 40 and 50 days after sowing of groundnut can effectively check the pest.
- ☆ Conserve the bio control population of spiders, long horned grasshoppers, preying mantids, robber fly, ants, green lace wing, damsel flies/dragon flies, flower
- ☆ Bugs, shield bugs, lady bird beetles, ground beetle, predatory cricket, braconids, trichogrammatids, NPV, green muscardine fungus.
- ☆ Spray any one of the following insecticide: quinolphos 2ml or chloropyriphos 3ml/lit of water

### C. Groundnut Leaf Miner *Aproaerema modicella*

#### Symptoms of Damage

- ☆ Young larvae initially mine into the leaflets, feed on the mesophyll and form small brown blotches on the leaf.
- ☆ Later stages larvae web the leaflets together and feed on them, remaining within the folds.
- ☆ Severely attacked field looks "burnt" from a distance.

#### *Management*

- ☆ Collect and destroy egg masses and early instars larvae
- ☆ Install light trap @ 12/ha for mass trapping.
- ☆ Release *Trichogramma chilonis* @ 50000/ha twice (7-10 days interval).
- ☆ Conserve the bio control population of spiders, long horned grasshoppers, preying mantids, robber fly, ants, green lace wing, damsel flies/dragon

flies, flower bugs, shield bugs, lady bird beetles, ground beetle, predatory cricket, braconids, trichogrammatids, NPV, green muscardine fungus

- Effective control could be achieved if insecticide is applied at 45 and 70 days after planting.
- Spray quinalphos 25 EC 2ml or methyldemeton 25 EC 1.6ml or dimethoate 30 EC 2ml per litre of water.

## Pigeon Pea

### A. Plume Moth *Exelastis atomosa*

#### *Symptoms of Damage*

- The larvae bore into the buds, flowers and pods and feed on the developing grains.
- By eating the buds holes are formed, and the larvae proceed by further eating up the grains.
- The larvae excretes inside the damaged grain sand pods and due to this fungus develops in it.

#### *Management*

- Spraying of the suitable insecticides should be done at 50 per cent flowering stage to protect the crop from the moth and larva.
- Spray monocrotophos 36 SL at the rate of 1l/ha.
- Quinolphos 2 ml/liter of water or Cypermethrin 1 ml/liter of water

### B. Blue Butterflies, *Lampides boeticus*

#### *Symptoms of Damage*

- Larvae chew leaves, buds, flowers, and pods

#### *Management*

- Spraying of 0.04 per cent monocrotophos (1 ml of 36 Sl/litre of water) or chlorpyriphos 20 EC @ 3.5 ml/lit. of water at 600-1000 lit. of spray material per ha.
- Spraying of cypermethrin 0.006 per cent or fenvalrate 0.006 per cent

### C. Blister Beetle *Mylabris pustulata*

#### *Symptoms of Damage*

- Feed heavily on flower and caused flower drops.
- Poor pod formation.

#### *Management*

- Manually picking them by hand or collecting them with an insect net and crushing them without touching them with hand.
- Cultivate less susceptible cultivar of pigeon pea

- ☆ Spray cypermethrin @ 30 g a.i./ha or permethrin @ 90 g a.i./ha at flowering stage.

### D. Spotted Pod Borer *Maruca testulalis*

#### *Symptoms of Damage*

- ☆ Larvae feed from inside a webbed mass of leaves,buds, and pods. Thus, Maruca is both a foliage feeder and pod borer.

#### *Management*

- ☆ Spraying of 0.04 per cent monocrotophos (1 ml of 36 Sl/litre of water) or chlorpyriphos 20 EC @ 3.5 ml/lit. of water at 600-1000 lit. of spray material per ha.

### E. Gram Caterpillar *Helicoverpa armigera*

#### *Symptoms of Damage*

- ☆ Bore holes on the buds, flower or pods
- ☆ Infested pods and flowers are webbed together.
- ☆ Caterpillars cause defoliation of the crop- all the leaves eaten away leaving the main stem alone

#### *Management*

- ☆ Deep summer ploughing.
- ☆ Intercrop one rows of red gram for every 5 or 6 rows
- ☆ Install pheromone trap @ 5/ha
- ☆ Use *Trichogramma chilonis* @ 1 lakh/ha or *Chrysoperla carnea* @ 50000/ha at 40 and 50 days after sowing of groundnut can effectively check the pest.
- ☆ Conserve the bio control population of spiders, long horned grasshoppers, preying mantids, robber fly, ants, green lace wing, damsel flies/dragon flies, flower bugs, shield bugs, lady bird beetles, ground beetle, predatory cricket, braconids, trichogrammatids, NPV, green muscardine fungus.
- ☆ Spray any one of the following insecticide: quinolphos 2ml or chloropyriphos 3ml/lit of water
- ☆ Apply fenvalerate (0.01 per cent) once at the time of pod setting or Monocrotophos (0.04 per cent).

### F. Pod Boring Weevil *Apion clavipes*

#### *Symptoms of Damage*

- ☆ Grubs enter the seed and start feeding on the developing grain and the whole grains is eaten.
- ☆ On young leaves it makes small holes and nibbling in flower buds.

#### *Management*

- ☆ Application of monocrotophos 40 EC (0.04 per cent) or Dipel 8L

## Mustard

### A. Mustard Aphid *Lipaphis erysimi*

#### *Symptoms of Damage*

- ☆ Both nymphs and adults suck the sap from leaves, buds and pods.
- ☆ Curling may occur in infested leaves and at advanced stage plants may wither and die.
- ☆ Plants remain stunted and sooty molds grow on the honey dew excreted by the insects.
- ☆ The infected fields looks sickly and blighted in appearance.

#### *Management*

- ☆ Destroy the affected parts along with aphid population in the initial stage.
- ☆ Early sowing of crop(mid October-November) escapes the aphid incidence.
- ☆ Conserve the following natural enemies: Ladybird beetles *viz.*, Coccinella septempunctata, Menochilus sexmaculata, *Hippodamia variegata* and *Cheilomones vicina* are most effcient pradators of the mustard aphid. Adult beetles may feed on an average of 10 to 15 adults/day.
- ☆ Spray the crop with one of the following in the flowering stage; oxydemeton methyl, dimethoate@ 625 - 1000 ml per ha.
- ☆ Two sprays of monocrotophos 0.04 per cent or phosphamidon 0.05 per cent one at 30 DAS and another at 45 DAS.

### B. Blue Butter Flies, *Lampides boeticus*

Same as pigeon pea

### C. Mustard Saw Fly *Athalia lugens proxima*

#### *Symptoms of Damage*

- ☆ Initially the larva nibbles leaves, later it feeds from the margins towards the midrib.
- ☆ The grubs cause numerous shot holes and even riddled the entire leaves by voracious feeding.
- ☆ They devour the epidermis of the shoot, resulting in drying up of seedlings and failure to bear seeds in older plants.

#### *Management*

- ☆ One spray of carbaryl (0.1 per cent) or Endosulfan (0.07 per cent) at 45 DAS

### D. Cabbage Butterfles *Pieris brassicae* and *Pieris canidia*

#### *Symptoms of Damage*

- ☆ The larvae butterflies feed voraciously on the leaves, branches and pods of the crop. The plants are defoliated with the result the small plants die while the grown up plants suffer in growth and yield.

***Management***

- ✰ The caterpillars in early stage should be controlled by hand picking and killing.
- ✰ Spraying with Malathion 50 EC at the rate of 1 litre per hectare controls the pest. In case there is attack of other pests also, use of Nuvacron 40 EC at the rate of 1 litre per hectare in 1000 litres of water has been found very effective.

## IFS Models for different Ecosystems of Manipur

### Module II (Livestock Feed Sources)

| Sl.No. | Ecosystems | IFS Components |
|---|---|---|
| 1. | **Valley** | i. Crop + Fish + Mushroom+ Horticulture |
| | | ii. Crop + Fish + water management + Duckery + Apiary |
| | | iii.Crop + Goat + Poultry + Horticulture |
| | | iv. Crop + Fish + Piggery+ Horticulture |
| | | v. Crop + Fish + water management |
| | | vi. Crop + Fish + Piggery + Horticulture + Apiculture + Mushroom |
| 2. | **Hill** | i. Crop + Fish + Horticulture + Piggery |
| | | ii. Crop + Fish + Duckery/Poultry + Agro forestry |
| | | iii. Crop + Livestock + Fish + Horticulture |
| | | iv. Crop + Fish + Horticulture + Apiculture + Agro forestry + Livestock |
| 3. | **Flood prone** | i. Crop + Fish + Horticulture + Duckery |
| | **areas** | ii. Crop + Fish + Duckery + Agro forestry |
| | | iii. Crop + Fish + Horticulture + Apiculture + Agro forestry + Livestock |
| 4. | **Jhum** | Crop + Agro forestry + Livestock |
| | | Crop + Agroforestry + Livestock + Horticulture + Apiculture + Allied enterprises |

### Modue II (Livestock Feed Management)

There are various types of feeding stuffs available for the livestock feeding. These feedstuffs can be grouped into different classes on the basis of bulkiness and chemical composition. However, no two feeding stuff are alike in the composition and characteristic but in practical feeding, substitution of one feeding stuff is made with another depending upon the market price and availability in a particular region. Therefore, it is necessary to know the categories of the feeds within which substitution are justified for the feeds having similar nutritional properties.

### Roughages and Concentrates

The first broad grouping of feeding stuffs is based on the bulkiness which depends on the content of crude fibre present. Roughages are the feeding stuff which are bulky and contain more than 18 per cent of crude fibre. They are relatively low in concentration of energy yielding nutrients. The roughages are straw, stovers, green

**Ecosystem 1: Valley ecosystem**

| Systems | Crop/Crop Sequence | Livestock | Fish | Fruits | Agro Forestry | Vegetables | Allied Enterprise | Water Management |
|---|---|---|---|---|---|---|---|---|
| **A. For small and marginal farmers (integration of all enterprises at small scale)** | | | | | | | | |
| i. Crop-fish- Allied enterprises | **Kharif:** Rice<br>**Rabi:** Mustard, Pea | | | | | | Mushroom/ Apiculture | Jalkund |
| ii. Fish-Duckery- Allied enterprises | | Duck | Carp or mixed culture | | | | Mushroom/ Apiculture | Farm pond/ Jalkund |
| iii. Crop-Fish-Duckery- Allied enterprises | **Kharif:** Rice<br>**Rabi:** Mustard, Pea | Duck | Carp culture | | | **Kharif:** Seasonal vegetables<br>**Rabi:** Seasonal vegetables | Mushroom/ Apiculture | Jalkund |
| iv. Crop-Fish-Piggery- Vegetables | **Kharif:** Rice<br>**Rabi:** Mustard, Pea | Cross breed | Carp culture | | | **Kharif:** Seasonal vegetables<br>**Rabi:** Seasonal vegetables | | Jalkund |
| v. Crop-Fish-Piggery- Poultry-Horticulture | **Kharif:** Rice<br>**Rabi:** Mustard, Pea | **Piggery:** Cross breed<br>**Poultry:** Vanraja, Gramapriya | Carp culture | Papaya/ Banana/ Citrus | | **Kharif:** Seasonal vegetables<br>**Rabi:** Seasonal vegetables | | Jalkund |
| **B. For medium and large farmers (integration of all enterprises at large scale)** | | | | | | | | |
| i. Crop + fish + duck + vegetables/fruits | **Kharif:** Rice/ Maize or both Soybean/ Groundnut<br>**Rabi:** Mustard, Pea, Potato, rabi maize | Duck | Carp or mixed culture | Papaya, Banana, Citrus, Pomegranate | | **Kharif:** Seasonal vegetables<br>**Rabi:** Seasonal vegetables | | Farm pond/ Jalkund |

Contd...

***Contd...***

| Systems | Crop/Crop Sequence | Livestock | Fish | Fruits | Agro Forestry | Vegetables | Allied Enterprise | Water Management |
|---|---|---|---|---|---|---|---|---|
| ii. Crop + fish + Poultry + vegetables/fruits | **Kharif:** Rice/ Maize or both Soybean/ Groundnut **Rabi:** Mustard, Pea, Potato, rabi maize | **Poultry:** Vanraja, Gramapriya | Carp or mixed culture | Papaya, Banana, Citrus, Pomegranate | | **Kharif:** Seasonal vegetables **Rabi:** Seasonal vegetables | | Farm pond/ Jalkund |
| iii. Crop + fish + Poultry + vegetables + fruits + Allied enterprises | **Kharif:** Rice/ Maize or both Soybean/ Groundnut **Rabi:** Mustard, Pea, Potato, rabi maize | **Poultry:** Vanraja, Gramapriya | Carp or mixed culture | Papaya, Banana, Citrus, Pomegranate | | **Kharif:** Seasonal vegetables **Rabi:** Seasonal vegetables | Mushroom/ Apiculture | Farm pond/ Jalkund |
| iv. Crop + fish + Piggery/Goatery + Poultry + vegetables + fruits | **Kharif:** Rice/ Maize or both Soybean/ Groundnut **Rabi:** Mustard, Pea, Potato, rabi maize | **Piggery:** Cross breed **Poultry:** Vanraja, Gramapriya | Carp or mixed culture | Papaya, Banana, Citrus, Pomegranate | | **Kharif:** Seasonal vegetables **Rabi:** Seasonal Vegetables(can also grow under poly house) | | Farm pond/ Jalkund |
| v. Crop + fish + Piggery/Goatery + Poultry + vegetables + fruits + Allied enterprises | **Kharif:** Rice/ Maize or both Soybean/ Groundnut **Rabi:** Mustard, Pea, Potato, rabi maize | **Piggery:** Cross breed **Poultry:** Vanraja, Gramapriya | Carp or mixed culture | Papaya, Banana, Citrus, Pomegranate | | **Kharif:** Seasonal vegetables **Rabi:** Seasonal Vegetables(can also grow under poly house) | Mushroom/ Apiculture | Farm pond/ Jalkund |

*Contd...*

**Contd...**

| Systems | Crop/Crop Sequence | Livestock | Fish | Fruits | Agro Forestry | Vegetables | Allied Enterprise | Water Management |
|---|---|---|---|---|---|---|---|---|
| vi. Crop + Goatery + Poultry + vege- tables + fruits/ Agroforestry + Allied enterprises | **Kharif:** Rice/ Maize or both Soybean/ Groundnut **Rabi:** Mustard, Pea, Potato, rabi maize | **Goatery:** **Poultry:** Vanraja, Gramapriya | | Papaya, Banana, Citrus, Pomegranate | Tree bean/ Fodder crops | **Kharif:** Seasonal vegetables **Rabi:** Seasonal vegetables(can also grow under poly house) | Mushroom/ Apiculture | Farm pond/ Jalkund |
| vii. Crop + fish + Dairy + Poultry + vege- tables + fruits + Allied enterprises + Fodder | **Kharif:** Rice/ Maize or both Soybean/ Groundnut **Rabi:** Mustard, Pea, Potato, rabi maize **Fodder:** | **Dairy:** Buffalo/cattle or both **Poultry:** Vanraja, Gramapriya | Carp or mixed culture | Papaya, Banana, Citrus, Pomegranate | | **Kharif:** Seasonal vegetables **Rabi:** Seasonal vegetables (can also grow under poly house) | | Farm pond/ Jalkund |
| viii. Crop + fish + Dairy + Poultry + vege- tables + fruits + Allied enterprises + Fodder + Allied enterprises | **Kharif:** Rice/ Maize or both Soybean/ Groundnut **Rabi:** Mustard, Pea, Potato, rabi maize | **Dairy:** Buffalo/cattle or both **Poultry:** Vanraja, Gramapriya | Carp or mixed culture | Papaya, Banana, Citrus, Pomegranate | | **Kharif:** Seasonal vegetables **Rabi:** Seasonal vegetables (can also grow under poly house) | Mushroom/ Apiculture | Farm pond/ Jalkund |
| ix. Crop + fish + Dairy + Poultry + vege tables + fruits + Allied enterprises + Fodder + Allied enterprises + Agroforestry | **Kharif:** Rice/ Maize or both Soybean/ Groundnut **Rabi:** Mustard, Pea, Potato, rabi maize | **Dairy:** Buffalo/cattle or both **Poultry:** Vanraja, Gramapriya | Carp or mixed culture | Papaya, Banana, Citrus, Pomegranate | Tree bean | **Kharif:** Seasonal vegetables **Rabi:** Seasonal vegetables (can also grow under poly house) | Mushroom/ Apiculture | Farm pond/ Jalkund |

*Contd...*

***Contd...***

| *Systems* | *Crop/Crop Sequence* | *Livestock* | *Fish* | *Fruits* | *Agro Forestry* | *Vegetables* | *Allied Enterprise* | *Water Management* |
|---|---|---|---|---|---|---|---|---|
| | | | | **Low lying areas (Flooded)** | | | | |
| Makhana/Chestnut + Fish + dairy/Piggery + vegetables/fruits | **Deep-water:** rice/maize/ Pulses before flood. During rabi season sweet potato, rabi maize, mustard, pea etc. | **Dairy:** Buffalo/cattle or both **Piggery:** Cross breed | Air breathing/ carp | Water-chestnut, makhana, mango, Water melon | | On bunds seasonal vegetables | | |
| Crop Makhana/ Chestnut +Fish + piggery/goatery + vegetables/fruits | **Deep-water:** rice/maize/ Pulses before flood, during rabi season sweet potato, rabi maize, mustard, pea etc. | **Piggery:** Cross breed | Airbreathing/ carp | Water-chestnut, makhana, mango, Water melon | Tree bean in surroun-ding areas | On bunds seasonal vegetables | | |
| | | | | **Ecosystem 2: Hill ecosystem** | | | | |
| | | | **A. For small and marginal farmers (integration of all enterprises at small scale)** | | | | | |
| i. Crop-fish-Allied enterprises | **Kharif:** Upland rice, maize, soybean, Groundnut, Beans **Rabi:** Mustard, Pea, Potato | | Carp or mixed culture | | | | Mushroom/ Apiculture | Jalkund |

*Contd...*

*Contd...*

| Systems | Crop/Crop Sequence | Livestock | Fish | Fruits | Agro Forestry | Vegetables | Allied Enterprise | Water Management |
|---|---|---|---|---|---|---|---|---|
| ii. Crop-Fish-Fruits-Vegetables | **Kharif:** Upland rice, maize, soybean, Groundnut, Beans **Rabi:** Mustard, Pea, Potato | | Carp or mixed culture | Banana, Papaya, Citrus, Passion fruits, Orange, Citrus, Colocasia, Yam, Jackfruit, Guava | | **Kharif:** Seasonal vegetables **Rabi:** Seasonal Vegetables (can also grow under poly house) **Spices:** Zinger/Turmeric | Mushroom/ Apiculture | Farm pond/ Jalkund |
| iii. Crop-Fish-Poultry-Agroforestry-Allied enterprises | **Kharif:** Upland rice, maize, soybean, Groundnut, Beans **Rabi:** Mustard, Pea, Potato | **Poultry:** Vanraja, Gramapriya | Carp or mixed culture | | Tree bean, fodder, Subabul | **Kharif:** Seasonal vegetables **Rabi:** Seasonal vegetables (can also grow under poly house) **Spices** Zinger/Turmeric | Mushroom/ Apiculture | Jalkund |
| iv. Crop-Fish-Piggery-Vegetables | **Kharif:** Upland rice, maize, soybean, Groundnut, Beans **Rabi:** Mustard, Pea, Potato | Cross breed | Carp culture | | | **Kharif:** Seasonal vegetables **Rabi:** Seasonal vegetables (can also grow under poly house) **Spices** Zinger/Turmeric | | Jalkund |

*Contd...*

***Contd...***

| *Systems* | *Crop/Crop Sequence* | *Livestock* | *Fish* | *Fruits* | *Agro Forestry* | *Vegetables* | *Allied Enterprise* | *Water Management* |
|---|---|---|---|---|---|---|---|---|
| v. Crop-Fish-Piggery-Poultry/Duckery-Horticulture-Allied enterprises | **Kharif:** Upland rice, maize, soybean, Groundnut, Beans **Rabi:** Mustard, Pea, Potato | **Piggery:** Cross breed **Poultry:** Vanraja, Gramapriya | Carp culture | Banana, Papaya, Citrus, Passion fruits, Orange, Citrus, Colocasia, Yam | Tree bean, fodder, Subabul | **Kharif:** Seasonal vegetables **Rabi:** Seasonal vegetables (can also grow under poly house) **Spices** Zinger/Turmeric | Mushroom/ Apiculture | Jalkund |
| **B. For medium and large farmers (integration of all enterprises at large scale)** | | | | | | | | |
| i. Crop + fish + duck/ poultry+vegetables/ fruits | **Kharif:** Upland rice, maize, soybean, Groundnut, Beans **Rabi:** Mustard, Pea, Potato | Duck **Poultry:** Vanraja, Gramapriya | Carp or mixed culture | Banana, Papaya, Citrus, Passion fruits, Orange, Citrus, Colocasia, Yam, Pomegranate, Grapes | | **Kharif:** Seasonal vegetables **Rabi:** Seasonal vegetables (can also grow under poly house) **Spices** Zinger/Turmeric | | Farm pond/ Jalkund |
| ii. Crop + fish + Poultry + vegetables + fruits + Allied enterprises | **Kharif:** Upland rice, maize, soybean, Groundnut, Beans **Rabi:** Mustard, Pea, Potato, rabi maize | **Poultry:** Vanraja, Gramapriya | Carp or mixed culture | Banana, Papaya, Citrus, Passion fruits, Orange, Citrus, Colocasia, Yam, Pomegranate, Grapes | | **Kharif:** Seasonal vegetables **Rabi:** Seasonal vegetables (can also grow under poly house) **Spices** Zinger/Turmeric | Mushroom/ Apiculture | Farm pond/ Jalkund |

*Contd...*

**Contd...**

| Systems | Crop/Crop Sequence | Livestock | Fish | Fruits | Agro Forestry | Vegetables | Allied Enterprise | Water Management |
|---|---|---|---|---|---|---|---|---|
| v. Crop + fish + Piggery/Goatery + Poultry + vegetables + fruits +Agro forestry + Allied enterprises | **Kharif:** Upland rice, maize, soybean, groundnut, Beans **Rabi:** Mustard, Pea, Potato, rabi maize | **Piggery:** Cross breed **Poultry:** Vanraja, Gramapriya | Carp or mixed culture | Banana, Papaya, Citrus, Passion fruits, Orange, Citrus, Colocasia, Yam, Pomegranate, Grapes | Tree bean, fodder, Subabul | **Kharif:** Seasonal vegetables **Rabi:** Seasonal vegetables (can also grow under poly house) **Spices** Zinger/Turmeric | Mushroom/ Apiculture | Farm pond/ Jalkund |
| vi. Crop + fish + Dairy + Poultry + vegetables + fruits + Agroforestry + Allied enterprises + Fodder | **Kharif:** Rice/ Maize or both Soybean/ Groundnut **Rabi:** Mustard, Pea, Potato, rabi maize **Fodder:** | **Dairy:** Buffalo/ Cattle/Mithun **Poultry:** Vanraja, Gramapriya | Carp or mixed culture | Banana, Papaya, Citrus, Passion fruits, Orange, Citrus, Colocasia, Yam, Pomegranate, Grapes | Tree bean, fodder, Subabul | **Kharif:** Seasonal vegetables **Rabi:** Seasonal vegetables (can also grow under poly house) **Spices** Zinger/Turmeric | Mushroom/ Apiculture | Farm pond/ Jalkund |
| **Ecosystem 3: Flood prone ecosystem** | | | | | | | | |
| **Low lying areas (Flooded)** | | | | | | | | |
| Makhana/Chestnut+ Fish + Duckery/ Piggery + vegetables/fruits | **Deep-water:** rice/maize/ Pulses before flood, during rabi season sweet potato, rabi maize, mustard, pea etc. | Duckery **Piggery:** Cross breed | Air breathing/ carp | Water-chestnut, makhana, mango, Water melon | | On bunds seasonal vegetables | Bee keeping, Mushroom and value added products | Rain water harvesting through Jalkund and micro rain water harvesting |

*Contd...*

***Contd...***

| Systems | Crop/Crop Sequence | Livestock | Fish | Fruits | Agro Forestry | Vegetables | Allied Enterprise | Water Management |
|---|---|---|---|---|---|---|---|---|
| Crop Makhana/ Chestnut+Fish + piggery/goatery + vegetables/fruits | **Deep-water:** rice/maize/ Pulses before flood, During rabi season sweet potato, rabi maize, mustard, pea etc. | | Air breathing/ carp | Water-chestnut, makhana, mango, Water melon | Tree bean in surrounding areas | On bunds seasonal vegetables | Bee keeping, Mushroom and value added products | Rain water harvesting through Jalkund and micro rain water harvesting |
| | | | **Ecosystem 4: Jhum ecosystem** | | | | | |
| | | | **Low lying areas (Flooded)** | | | | | |
| Crop + Agroforestry + Livestock | **Kharif:** Upland rice and Maize with groundnut/ soybean either sole or intercrop **Rabi:** Pea, other pulses | **Piggery:** Cross breed | | | Tree bean and other leguminous trees in surrounding areas | | | Farm pond/ Jalkund |
| Crop + Agroforestry + Livestock + Allied enterprises | **Kharif:** Upland rice and Maize with groundnut/ soybean either sole or intercrop **Rabi:** Pea, other pulses | **Piggery:** Cross breed **Poultry:** Vanraja, Gramapriya | | | Tree bean and other leguminous trees in surrounding areas | | Mushroom/ Apiculture | Farm pond/ Jalkund |

*Contd...*

***Contd...***

| *Systems* | *Crop/Crop Sequence* | *Livestock* | *Fish* | *Fruits* | *Agro Forestry* | *Vegetables* | *Allied Enterprise* | *Water Management* |
|---|---|---|---|---|---|---|---|---|
| Crop + Agroforestry + Livestock + Fish + Horticulture + Allied enterprises | **Kharif:** Upland rice and Maize with groundnut/ soybean either sole or intercrop **Rabi:** Pea, other pulses | **Piggery:** Cross breed **Poultry:** Vanraja, Gramapriya | Carp or mixed culture | Banana, Papaya, Citrus, Pomegranate | Tree bean, fodder, Subabul | **Kharif:** Seasonal vegetables **Rabi:** Seasonal vegetables | Mushroom/ Apiculture | Farm pond/ Jalkund |

fodders, hay, silages, etc. Concentrates are less bulky and therefore contain less crude fibre (below 18 per cent). They are rich in energy yielding nutrients. Under concentrates are the cereal grains, oilseeds and their by products and animal products. Roughages cannot be substituted for the concentrates.

The roughages maybe dry or green. They can also be grouped into maintenance quality containing about 3 to 5 per cent DCP, non maintenance quality containing more than 5 per cent DCP on dry matter basis. The maintenance type roughages are growing maize, oat, etc. The non maintenance types are straws, hays, etc. The productive type roughages are cowpea, lucerne, berseem etc. The term roughages is also used for nutritionally adequate roughage. However, most of the people used the term forage, fodder and roughage interchangeably. The concentrates are broadly divided into three categories. In the first category, those concentrates comes which are rich in energy and are low in DCP. The examples are the cereals grains like wheat, maize, sorghum, etc. In the second category are the concentrates which are very rich in DCP like groundnut cake and mustard cake. They are also called proteinous concentrates. The third category is in between where protein and energy are in medium range like rice bran, wheat bran and *chunies*. The concentrates are also from the animal sources like fish meal, meat meal, blood meal, etc.

### Forages

The roughages maybe dry or green. The roughages are either available as by product from the main cereal crop like wheat, straw, paddy straw or maybe specially prepared from the green fodders like preparation of hay. Here, only some of the fodders which are available in Manipur along with some nutritious fodders which are not yet cultivated but have a huge potential for fodder cultivation as a source of energy and protein will be discussed.

## Fodder Crops

### 1. Kharif Fodders

a. Cowpea (*Vigna sinensis*): Cowpea is important, quick growing legume, which fits well in crop rotation. It is adopted to warm summer climate and can be grown on all types of soils ranging from sandy loam to heavy loams soil, if they are well drained. It is susceptible to water logging conditions. It can be sown alone or with non-leguminous like maize.

b. Maize (*Zea mays*): It is one of the most important crop and also principal kharif crop in Manipur. It requires warm and temperate climate and grows well on alluvial loam, fertile and well drained soil. The crops grow well in slightly acidic to neutral soils (pH 5.5- 7.5). It can be harvested in the pre-flowering stage for use as a green fodder because in this period the crude fibre content is less but the DCP is more. It is a maintenance type of fodder containing 8-10 per cent of protein. It can be used as dual purpose crop like cowpea. When the cobs is milk stage are removed the residual plant constitute a maintenance ration. When however, the crop is allowed to mature the stalk becomes very hard and is not even comparable with the paddy straw.

Cowpea

Sorghum

Maize and Cowpea Intercropping

Bajra

c. Teosinte (*Eucularea Mexicana*): Teosinte is a close relative of cultivated maize. It is a kharif crop and requires hot and temperate climate. It required rich and wel drained soil for good growth. The plant tiller profusely and give rise to a number of shoots forming thick clumps. The fodder is ready to use in the period of 90-100 days. The yield of fodder is about 350 to 500 quintals per hectare. It is maintenance quality fodder.

d. Sorghum (*Sorghum vulgare*): It is one of the most important kharif fodders grown in many parts of our country and has a huge potential for fodder in Manipur. It can grow well in irrigated and rainfed conditions. It does well under hot and dry climate. It requires well drained soil and appears susceptible to water logging conditions. Though it can be grown in any type of soil but loam soil have proved to give more yield. The sowing can be done from April onwards upto July. The fodder can be harvested within 70-90 days with a yield of 250-450 quintals per hectare depending on the variety.

e. Bajra (*Pennisetum typhoideum*): It is a quick growing disease resistant, high tillering crop suited to arid and semi-arid region. It grows well in sandy loam to clay loam soil. It grow throughout the country, it can withstand drought for a long period. It can be sown early in the spring under irrigated conditions and in kharif under rain fed conditions. It is not suitable in high rainfall areas.

## 2. Rabi Fodders

The important cultivated rabi fodders are oat, lucerne, berseem, Chinese cabbage, turnip, etc. Out of these, lucerne is a perinneal and stays in the field for 4-5 years.

a. Oats (*Avena sativa*): Oats is a cheap irrigated winter fodder for livestock. It comes up well in temperate region, in sandy loam to clay loam soils. It can be sown in the month of last September to November. It requires frequent irrigation. Early varieties are harvested in 110 days and the late one in 130 days. First cutting gives 25 tonnes per hectare and second cutting 20 tonnes per hectare. Oats maybe considered as maintenance quality fodder containing about 7-9 per cent crude protein. The content of crude protein can be increases upto 11 per cent by nitrogen fertilizers in which case it may be considered as productive fodder.

b. Berseem or Egyptian Clover (*Trifolium alexandrium*): It is one of the most important fodder crops and has rightly been described as the king of fodders. It is highly esteemed fodder which has a special place in animal husbandry programmes throughout the country. Though a migrant from Egypt it has established in these region for the last 60 years and may well be considered as a native of India. It has many desirable qualities. Ten to fifteen kilograms of fodder alone with straw constitute a maintenance rations. It can support growth and milk production *adlib* feeding, balanced by straws. It does not tolerate acidic soil but grow in other kinds of soils except *usar* lands. The crop is ready in 55-60 days after sowing for the first cutting. Subsequent cutting are taken at 30 days interval. In all 5-6 cutting can be obtained. The total yield is 500-600 quintals per hectare. Some of the recommended varieities are; ordinary diploid and giant berseem.

c. Lucerne (*Medicago sativa*): It is a perinneal crop once grown it continue to supply nutritious fodder for 3-4 years. It is sown both as an annual and perinneal crop. The crop is generally grown in irrigated areas of dry tracks and gives yield in well drained drained sandy loam soil. This is a productive fodder which can support not only growth but milk production upto 8 kg when fed *adlib*. It can also support 8 kgs of milk yield in the cows and buffaloes when fed *adlib*. The best time of sowing is between October to end of November. First cutting is ready after 60 to 75 days from the date of sowing. If it is cultivated as annual fodder, the yield is 700-900 quintals per hectare. There are many varieties but type 8 and 9 developed at Fodder Research Station at Sirsa are recommended for adoption.

d. Mustard (*Brassica* spp.): It grows well in sandy loam soil. The sowing time is early September to November. It is sown just after the kharif crop. It is a short duration crop of 60-70 days. It is harvested at the flowering stage. The green fodder yield is 30-35 tonnes per hectare. It has a very high percentage of protein, so it should be fed along with straw to livestock's. Besides these, Chinese cabbage and turnip are also a good rabi crops for livestock feeding practices.

## 3. Perennial Grasses

a. Para Grass (*Brachiaria mutica*): It is also called as Buffalo grass. It grows best under swampy conditions. It is a rapid summer growing perennial spreading by surface runners which are often 6 m in length throws a broad hairy flag. When the field is completely covered the plant sends erect shoots which may be 1 to 1.5 m in length. It is propagated by roots and stem cuttings as seeds are not viable. It is ideally suited for heavy rainfall areas. It also grows well on boundaries of ponds, tanks or banks of ditches/ canals. It is very sturdy and aggressive grass and when once established propagate itself easily even under sewage irrigation. It is ready for first cutting around 90 days after the cultivation and subsequent cut can be taken once in 35 days. It is a palatable grass relished by the cattle.

b. Hybrid Napier (NB-21): It is a cross between napier grass and bajra. It is a tall perennial quick growing grass which is propagated vegetatively. It comes up well in loam and clay loam soil. It can be planted in the month of May to June. It requires irrigation once in 10 days in summer and in 20 days in other season. First cut can be done at 60 days and subsequent once in 35-40 days. The yield is 150 tonnes per hectare. The grass is suited for ensilage and should not be allowed to over mature.

c. Guinea Grass (*Panicum maximum*): It is a perennial which grows in tufts which may tillers arising from one stem. It grows well in medium or rich well drained loamy soil. It is planted in the month of March-April. Irrigation is needed once in 15-20 days. First cutting can be taken in 75 days and subsequent cutting at the interval of 45 days. In severe winter the grass becomes dormant. The fodder is best relished by cattle. The yield is 120 tonnes per hectare but under sewage irrigation heavy yield can be obtained.

d. Dinanath Grass (*Pennisetum pedicellatum*): It is an annually growing fodder, grows well in both rainfed and irrigated conditions. It tolerates slightly alkaline to acidic soil. It is useful for marginal wasteland. The seed germinates quickly and responds well in clay loam and sandy soil. It is sown in the month of June-August. Irrigation is needed once in 20 days in summer and 30 days post monsoon. Cutting can be made at 90 days and subsequent cutting at 60 days interval. The yield is around 75 tonnes per hectare per year. This grass is also relished by cattle.

Para Grass

Dinnanath Grass

Oats

Mustard

Guinea Grass

Napier NB-21

## 4. Tree Fodders

Tree leaves are useful as fodder because of their availability in lean season, cheapness, good nutritive value, requiring less maintenance and causing no

Subabul

Ficus Leaves

Jackfruit Leaves

infringement on cultivated lands. In this part of the region, there are many varieties of tree leaves which can be utilised as fodders during lean period. Some of the commonly available trees are: Subabul, *Acacia* sp., Jackfruit, Mulberry, Bamboo and *Ficus* sp.

## 5. Locally available Feed Sources

During scarcity of fodders during lean period, locally available plants are fed to the cattle, pigs etc to meet their demand. Besides *Colocasia sp* and sweet potato some of the locally available plants used as feeds are Banana stem, Chinlengbi (*Spilanthus acmela*), Sampakpi (*Bidens pilosa*) etc.

## Conservation of Fodder

Conservation of fodders and grasses refers to the preservation of these foods in such a manner that they can be stored for long periods without much loss in their nutritive value. Some the methods in use are:

Banana Tree

Chopped Banana Stem

Colocasia

Chinlengbi

Terapaibi

Local Plants Fed to Pigs

1. Preserving green forage as such, out of contact with air.
2. Natural drying in the sun.
3. Drying tender green grass in artificial driers.

Sweet Potato

Kitchen Waste Fed to Pigs

## Silage Making

Silage is a fermented green forage product prepared under an anaerobic condition. Good silage made from quality crops is yellow- brown in colour, with a characteristic acid fruity smell and just exude moisture when squeezed. It is highly palatable, relished by the cattle. It provides succulent feed at low expense all the year round. It can be prepared from plants having thick stems which are generally not suitable for hay making. It preserves the carotene of the forage. It requires less storage space than as dry forage, less expensive and time consuming in making than hay making.

## Silage Making Process

Of the two kinds of silo *i.e.* tower or trench, it is usually trench is cheap and easy to construct. Trench silo may be kacha or pucca or rectangular pits, narrower at the bottom than at the top to facilitate effective packing of silage material. The floor and the sides may vary from plain soil surface to line surfaces with stone slabs and brick. The floor is slightly sloppy to permit drainage. Bottom and side linings are done to reduce the wastage. Underground silos are cheap and easy to make. Round pits are preferable. Straw or dry grasses can be placed at the bottom and sides of the pit to prevent fodder from coming into contact with earth and getting damaged. Care is taken to construct on high and hard soil not subject to percolation of water.

The crop is first harvested at the optimum stage and wilted where ever necessary to get the desired dry matter content (35 per cent). The crop is chaffed to 2-4 cm in length, rapidly filled in silo spread in successive layers and tightly packed. Proper trampling is done by man, bullocks or tractor depending on the size of pits to compact the material. The pit is packed up to some 1.5m above the ground, the top surface packed with the long paddy straw, sugarcane leaves or poor quality grasses or polythene sheets to seal the material. It is than covered with earth and plastered with mud. While fodder settles cracks will appear in the covering layer which should repaired with mud. Enough sloped is given on top surface to protect against flowing rain water by a low bund all around.the finished pit will appear as a low mound. After about 90 days the silage will be ready for feeding. Only required quantity for feeding animal daily should be removed from the pit. A $m^3$ of silage weighs 650 kg (moderate packing) or 700 kg when tightly packed. Based on this, the size of the pit can be calculated. The forage crop from 0.3 ha in a circular silo 3m diameter and 2.5m depth may be enough to feed 5 animals at 20 kg/day for 150 days.

## Hay Making

Hay making is very common in our country with the farmers but unfortunately it is not properly made. Generally the grasses and fodders are harvested when the crops are mature and are not properly dried with the result that a hay of very poor quality is prepared.

To prepare good quality hay the crop should be harvested at pre-flowering stage so that maximum amount of nutrients is conserved.

It should be prepared in such a way that there is less loss of leaves due to shattering and maximum amount of green colour is retained by the hay. The hay should be free from mould,bad odour and free from weeds. It should be palatable and should have not more than 15 per cent moisture.

### Crops Suitable for Hay Making

The fodder crops having soft and pliable stem are more suitable for hay making. Green oat is the best crop for hay making. However green berseem, Lucerne, cowpea etc can also be used for hay making if proper care is taken in cutting whereby shattering of leaves is avoided. Perennial grasses like Napier, guinea and annual crops like maize, sorghum are suitable for hay making.

### Method of Hay Making

For the production of good quality hay the crop should be harvested early in the morning when the dew is dried. After cutting the crop is left as such for few hours for curing. After about 4-5 hours if there is a good sunshine the fodder is turned upside downside with the hay rake. In the evening small loose heaves can be formed with the hay rake and fodder is left. On the next day one or two turnings are given and by afternoon moisture level will come down to 25 per cent. At this stage it can be baled and kept in the bale form or if it is a heavy rain fall area it can be stored on tripod stand. This tripod system has an advantage that if there is a rain the water will passed and there is proper aeration from below. Various method of drying are practiced, drying the crop on wire, roof tops, tree tops, galvanised tin sheets, tripod stand etc. In all these cases care should be taken to avoid shattering of leaf in leguminous crop.

# Module III: Health and Disease Management of Swine in IFS

## Deficiency Diseases

Minerals and vitamins, in adequate quantities, are essential in the diet of all animals. Many diseases occur due to deficiency of minerals and vitamins. Minerals and vitamins are more necessary in pigs due to their higher growth rate. It is normally seen that when pigs are reared on self made rations, they suffer more with the deficiency of one or more mineral and vitamins. Thus, it is necessary to introduce extra mineral and vitamins in the pig ration. If it is not done, its growth rate and body weight do not increase in proper ratio. The excess and deficiency of one mineral depends on quantities of other minerals present in the ration. It is frequently seen that pigs suffer from deficiency of more than one mineral and vitamin but some time pigs suffer from deficiency of one substance and excess of the other substance, so a balanced ration – with appropriate proportions of various mineral and vitamins – is necessary. Many times it is seen that some infectious disease increase the necessity of mineral and vitamins in the body. This may cause deficiency of mineral and vitamins in the body and the animal may not recoup.

Following are the description of these symptoms and their prevention.

## Zinc Deficiency

Zinc deficiency is known as parakeratosis. Zinc deficient patients show peculiar symptoms in their skin and secrete a watery fluid which becomes dry after some time. There is loss of hair from the affected areas of skin with weight loss. Symptoms of zinc deficiency resemble those of scabies, but when zinc is mixed in the ration of these pigs, the problem is rectified rapidly. Zinc can be mixed in the ration @ 100 gms zinc per quintal (100 kgs) feed to remove the symptoms of zinc deficiency on the skin of affected pigs.

## Iron Deficiency

Iron is necessary for blood formation. Iron deficiency causes anaemia. Iron deficiencies frequently occur in piglets, where mother is also deficient in iron content at the time of pregnancy. This disease is known as piglet anaemia.

Newly born piglets are born piglets are deficient in iron because it does not cross placental barrier from mother to foetus and mother's milk is also deficient in iron. So, iron is needed by the piglets. This disease is dominant today due to management.

Anaemia due to deficiency of iron is characterized by whiteness of mucus membrane in piglets, dullness and separation from other piglets. Appetite is reduced with weight loss. To correct the iron deficiency, give 20 gms iron per kg ration to pregnant sow during the last two weeks of pregnancy. Therefore, newly born piglets may not have iron deficiency. Fifteen mg/day ferric choline citrate may be given till weaning. The amount may be given by the oral route or by injection. Newly born piglets may be given 4 mg of 1.8 per cent ferrous sulphate ($FeSO_4$) in injection form till 7 days after birth. To avoid injection everyday some veterinarians advise to give Ferrous Sulphate @0.5gm to 1.0 gm per week. Some persons (piggery man) apply a mixture of 450 gm FeSO4, 75 gm CuSO4, 45 gm sugar dissolved in two litres of water as a thin paste on the teeth of sow for many days. It fulfils the iron deficiency in piglets.

In India, to prevent piglet anaemia, give iron injections (1 mg Impheron or iron dextrin on the 3$^{rd}$ day and 14$^{th}$ day after farrowing.

### Copper Deficiency

Copper deficiency frequently occurs in piglets. When piglets attains the age of 5-8 weeks. Copper (Cu) deficiency may cause weakness in the hind quarters, piglets feels difficulty in movement and there is ataxia, If Cu deficiency is of the per acute type than pigs are unable to stand from hind trunk. Copper is essential for blood formation. Its deficiency also causes anaemia. To remove the deficiency of copper from pigs, mix 125-250 mgs copper sulphate per kg of the feeding ration. Often it is seen that pigs more than 50 kgs live body weight are less benefitted after mixing CuSO4 in their feeding ration. So, it is necessary to remove copper deficiency as soon as possible.

### Iodine Deficiency

Iodine is necessary for all round development of the body. It is necessary for completion of many reactions inside the body. Iodine deficiency in pig very frequently causes goitre. Some factors like nitrate, thiocyanate, glucocynolate when mixed in feeding ration causes iodine deficiency in the body and shows symptoms of iodine deficiency like the piglets being hair less (Alopecia). Many piglets die just after farrowing. Live piglets suffer with goitre, which can be seen externally. Iodine deficient piglets show weakness in legs and feel difficulty in locomotion.

To remove iodine deficiency, give 200 mg KI per kg body weight in the feeding ration of pregnant sows. Sometimes iodine deficiency is seen when pigs are fed with groundnut mixed rations. In this condition, ration changing is necessary if possible give 200-300 mg KI or 400- 500 mg KI injection, two times at one month interval to the pregnant sows.

Piglets ration are mixed with small quantities of KI (50mgs) per day. Paint iodine on the udder of the sows two times a day to save the piglets from iodine deficiency.

### Calcium Deficiency

When pig ration has calcium: phosphorus ratio of 1:1 or less, then pigs may show symptoms of Ca deficiency. In appetence, frequent constipation, weakness, weak bones and easy bone fracture are observed.

To remove these symptoms, give calcium in adequate quantities in the appropriate ratio of Ca and P in the pig's rations otherwise calcium will not be utilized properly.

### Selenium (Se) and Vitamin E Deficiency

Se deficiency in pigs causes mulberry heart disease. Selenium 7 vitamin e show metabolic similarities. Often, both show deficiencies simultaneously. If ration is mixed with selenium or vitamin E, mulberry heart disease symptoms disappear.

Mulberry heart disease patients (pigs) die suddenly. Post mortem reveals light coloured liver, thin muscles, spot on the liver and muscle fibres brown coloured fat. When above symptoms appear and heart muscles show some light colour spots, it is assumed that pigs are suffering from selenium deficiency.

Death stops after mixing Se or vitamin e in their ration. In addition, vitamin E is helpful in maintenance of reproductive performance in female pigs. Se and vitamin E also help in the formation of strong texture and lustre of the skin.

## Vitamin D Deficiency

Vitamin D deficiency more frequently occur in the piglets. It is known as rickets. In rickets, bents appear on the legs of the piglet. If not treated properly and timely, leg deformity may become permanent. Growing piglets require vitamin D in adequate quantities. Its deficiency also shows deficiency of Ca, P and some other mineral salts. These deficiencies cause weak bone and reduced growth rate. If vitamin D is not in adequate quantities, then certainly ca and P deficiencies will occur. So it is necessary to add adequate vitamin D in the ration of growing piglets and per day nearly for 30 minutes the piglets should be kept in the ultra violet sun light. In the presence of ultra violet rays, pig produce vitamin d. it is better to apply a massage of mustard oil on the legs of pig and they should be left free to move in the sunlight. With these, give vitamin d (300 i.U. per kg ration) and ca and P mixture for 15-20 days to prevent rickets.

## Vitamin A Deficiency

Vitamin A deficiency occurs either due to its deficiency in the ration or due to reduced absorption of vitamin A from the intestine. Their deficiencies in piglets cause defective brain development and in adults it causes night blindness. Skin becomes rough, eye pupil becomes white, there is loss of weight. Dystocia may occur in the pregnant animals. Four to five months old piglet shows severe symptoms.vitamin a deficiency should be corrected very fast otherwise clinical symptoms become permanent. In vitamin a deficiency, give 440 I.U. vitamin A/kg body weight in the form of injection. For quick relief, water mixed injection is used, and for longer reliefs, vitamin A is mixed with oily substance and then in ration. To prevent vitamin A deficiency permanently, mix retinol (0.3μg0 or 5-8 μg β-carotene per kg per day. Some piggeries feed 3000-6000 i.U./kg body weight vitamin A in the ration in 50-60 days interval.

## Vitamin B Group Deficiency

Vitamin B is a group of water soluble compounds. Vitamin B group compounds are synthesized in the digestive system. Because growth rate in pigs is very fast, so their deficiencies may appear in the pigs. Main vitamins of group B are thiamine, riboflavin, nicotinic acid or niacin, pyridoxine, cyanocobalamine, lipoic acid and benzoic acid their deficiency causes beri beri disease. Embryo death in uterus, weakness, ataxia, dry skin, black tongue, diarrhoea or constipation, loss of hair, reduced growth rate are observed.

As soon as symptoms of this vitamin B group deficiency appear, it may be treated with diet or injection of vitamin B group compounds otherwise permanent deformity occurs in the body. Injection of these group vitamins may be given on the 4$^{th}$ day of birth and the next injections may be given on the 4$^{th}$, 5$^{th}$, 15$^{th}$, 16$^{th}$ and 17$^{th}$ days. Vitamin B group are not stored in the body. Thus, give adequate amounts of vitamin B grouper per day.

## Niacin Deficiency

Pig is highly prone to niacin deficiency. if this vitamin is provided in less amounts in the ration of pigs, it reduces the growth rate and causes intestinal ulcers.

Niacin deficiency causes diarrhoea in pigs, loss of water from body and loss of weight. The symptoms of niacin deficiency resemble those of salmonellosis but antibiotic treatment does not cure the disease. So, they must be provided in balanced quantities by the feed. the growth rate of pigs body is fast and so, the importance of mineral and vitamins is more in pigs.

## Systemic Diseases

### (A) Diseases of the Digestive System

Function of digestive system is ingestion of feed, its digestion and absorption. Any disturbances in overall processes affect digestion and digestive system both and produce abnormalities in the body varying extents. The animal feels unhealthy and sometimes death may occur. Therefore, for a good piggery, the farmer should give due attention to the pigs. They must take normal feed and water. Their digestion and excretion should be proper. If it is not normal, then the veterinary officer may be contacted.

Following diseases are commonly found in the digestive system of pigs:

### 1. Gastritis

After taking feed, it passes from mouth and oesophagus and first raches the stomach. The stomach is in the form of a big beginning the stomach, gastric juice is secreted from its wall. This mixes with the feed. Some food remains undigested in the stomach. If any disturbance occurs in stomach, it influences the digestion of food.

***Causes***

1. Pigs feed on unfit or fungus infected feed.
2. Feeding of hard or sharp things like needle etc.
3. Feeding of some poisonous fungi, copper sulphate, led, arsenic etc.
4. Bacterial and viral infections, i.e pig diarrhoea, collibacilllosis, salmonellosis, swine fever etc.
5. Intestinal parasites like the round worm

**Symptoms:** on the basis of severity, the symptoms are divided into two parts:

1. **Acute gastritis**: beginning of vomiting. It may be bloody. Pigs refuse feed and exhibit polydipsia, abdominal pain, dehydration due to excessive vomition and rapid respiration.
2. **Subacute gastritis:** In appetence and vomiting occur just after taking feed. There is subacute abdominal pain and pigs become weak day by day.

Treatment: remove the cause of gastritis; if pig ingests any poisonous substance with feed then stomach lavage is done by suitable solution. If vomitus is more then give antiemetic i.e perinorm, sequil, avomine etc. And intravenous electrolyte solution is necessary. Kaolin, pectin, tetracycline etc. May be given by oral route. In village, if electrolyte or glucose solution are not available, then give 250-500 gms rabb, 10-20 gms salt, 0gm $CaCo_3$, 250 gm barley or rice cooked supernatant water (gruel) is helpful.

**2. Constipation**

constipation and impaction are disease of adult pigs. In constipation ball like hard faeces appear in large intestine, which does not pass towards outside and cause obstruction. Animal becomes dull, off feed and does not drink, faeces are dry, hard, and covered by mucus. In constipation, give $MgSo_4$ (100-300 gms) or salt solution (25-100gms) or castor oil (25-100ml). If this treatment is not useful then contact the veterinarian.

## I. Dermatological Ailment

A dermatological case not responding to apparently rational symptomatic therapy needs more detailed investigation. A proper investigation approach to a dermatologic case should include the following (1) History (2) physical examination (3) laboratory and other diagnostic techniques (haematology, biochemistry, biopsy and identification of various etiological agents responsible for the malady).

Skin disease can be mainly classified into the following categories:

1. **Bacterial skin disease (pyoderma):** these include mainly the surface pyoderma and deep pyoderma.
2. **Parasitic skin disease:** these may be due to ticks, lice, mites, fleas, flies, helminth parasites, protozoa, rickettsia.
3. **Fungal skin disease:** include ringworm (*i.e.* Microsporum spp., *Trichophyton* spp., *Cryptococcosis, Blastomycosis*.
4. **Viral skin disease:** foot and mouth isease
5. **Endocrine skin disease:** may include hypothyroidism, hyperadrenocorticism, male eminizing syndrome, testosterone-responsive dermatitis.
6. **Seborrhoea:** seborrhoeic syndrome
7. **Immunological skin diseases:** include the type-I, type – II, type – III and type- IV hypersensitivities; food allergy, autoimmune disorders.
8. **Physical skin disease:** like nasal dermatitis. Burns and frost bite also affect skin.
9. **Chemical skin disease** may be due to chemicals like strong acids and alkali.
10. **Timorous conditions of the skin:** like papilloma, basal cell tumour, sweat gland tumours.
11. **Nutritional skin disease:** may be a result of a deficiency of protein, essential fatty acids, vitamins(viamin A, Vitamin B, Vitamin E), zinc iodine
12. **Inherited skin diseases:** like some alopacias.
13. **Skin conditions due to poisoning:** these may be due to snake bite,bee sting and wasp.
14. **Miscellaneous and other conditions:** inspite of all these groups, there are certain skin diseases which are not included in any of these groups and are classified as skin diseases of unknown etiology or multiple etiology. Diseases under this group include the panniculitis, dermatosis, pigmentary disturbances of unknown etiology (lentigo, vitiligo).

**Pyoderma:** The main causative microorganism includes staphylococcus epidermitis, Corynebacterium sp., Clostridium sp.

Symptoms include hot spot, wet eczema, summer eczema, exudative purulent dermatitis and diagnosis may be done by history, physical examination, isolation of causative organism from the lesion. Antibiotic sensitivity test may be performed to identify the more effective antibiotic. Treatment may include local cleaning with medicines and based on the results of the antibiotic sensitivity tests, parenteral treatment with antibiotics may also be advocated with surgical drainage in some cases.

## Parasitic Skin Diseases

The symptoms may include pruritis, secondary alopecia, anaemia in heavy infestation and diagnosis may be done by identifying the parasitic either macroscopically or microscopically. Treatment may include educating the owner, for example the life cycle of the flea may be explained to the owner. Prednisolone, on alternate days, may be prescribed and the environment may be treated with insecticidal sprays and powders.

In ear mange include shaking of the head, stretching of ears and brown to blackish waxy powdery deposit in ears. Diagnosis may be made by history, clinical signs and removing waxy powdery deposit of the infected ear. To treat, insecticide spray may be used to kill the mite and in case of the presence of more wax, clean by putting lukewarm water containing a pinch of washing soda and if pus is present, then use systemic antibiotics.

Scabies due to sarcoptes sp. Is a highly contagious skin disease exhibiting thickening of skin with intense irritation, alopecia, bare patches on the skin, hyperkeratosis' and folding of the skin. Diagnosis may be done by history, clinical, identification of mites in the scrapings which are dissolved in 10 per cent KOH or 10 per cent NaOH and treatment includes hair clipping, insecticide spray (1:500 solution of asuntol, benzyl benzoate preparations) and herbal preparation having karanj oil, garlic extract, turmeric powder, lemon extract, camphor, onion extract may be applied.

In case of demodectic mange, the mites live in the hair follicles and sebaceous gland, which they multiply and grow leading to loss of hairs, thickening of the skin. Diagnosis may be made by history, clinical signs and presence of cigarette shaped mites in the skin scrapings and treatment includes clipping of hairs, bath with antiseptic solution like savlon, dettol. Broad spectrum antibiotic with antihistamine drugs like Avil may be prescribed.

## Fungal Skin Diseases

These may affect the skin, hair, nails and diagnosis of the fungus may be made by woods lamp examination, fungus culture, and biopsy. Under treatment, clip the lesions, griseofulvin at the dose rate of 15-20 mg/kg may be given, topical treatment includes miconazole and in some animals amphotericin B may be useful.

In fungal infections due to Microsporum sp. And Trichophyton sp. There are circular areas of hair loss with scaling and erythema, itching and irritation, lesions always spread from the centre to periphery indicating recovery from centre showing ring like lesions. Diagnosis is made by KOH hair in 10 per cent KOH shows sheath of small spores in mosaic form which completely surround the hair at its base. To treat, Griseofulvin 25-50 mg/kg body weight is given generally treatment is given for one month and local application with 2-5 per cent boric acid or 6 per cent benzoic acid may be made.

## Eczema

Is a superficial skin inflammation which may be produced by various external (antiseptics, disinfectants, bedding, dyes) and internal (food allergens, protozoan allergen, bacterial allergen – Tuberculosis, leptospirosis, viral allergen; fungal allergen -Trichophytosis) substance against which cells are sensitized, but it always involves a special local or general predisposition. Initially there is a small reddish spot, swelling of superficial layer of skin and later there may be exudation of serum or pus. To treat remove the allergen, diet must readily digestible and should have all vitamins and minerals. Antihistamines, steroids and antibiotics are given to prevent secondary bacterial infection.

## II. Reproductive Disorders

The sow gives birth to 10-12 piglets at one stretch. Giving birth to many young ones at one moment makes it prone to problems of the female reproductive tract.

The main problems during pregnancy are:

### (1) Prolapse of the Uterus

The problem is common during the last phase of pregnantcy in which along with vagina and vulva, the cervix and uterus also bulge out. The main causes include mineral (calcium) deficiency and chronic infections of the female reproductive tract. It can be corrected manually. The hands and the uterus may be properly washed with potassium permanganate solution in water (1: 1000).first the uterus may be brought to its original position after which the front portion of the animal may be put on lower platform and the hind portion may be elevated around 6 inches from the ground level. Easily digestible diet should be given. Calcium and phosphorus containing medicines may be given orally or intravenously. If greater portions of the uterus have come out, then a ring of cloth may be tied around the mouth of the vagina with a thread. This thread may be tied to a thread whose other end may be tied to a thread tied around the abdomen.

Purse string sutures may be applied to the vagina. Sulpha drugs or antibiotics will be favoured.

### (2) Mummification

This problem may occur in one or two foetuses in which intra uterine death occurs and it dries because of absorption of water.the death foetusare also expelled out along with the live ones during farrowing. This occurs due to obstruction in the flow of nutrition in placenta or due to microbial infections of the reproductive tract or the foetus. Under therapy, the uterus may be cleaned and antibiotics may be advocated.

### (3) Expulsion of the Foetus (Abortion)

In this, dead and immature foetus are expelled out by the pregnant sow. The main causes include.

i. **Bacterial causes**; leptospira, brucella, streptococcus, corynebacterium, salmonella etc.

ii. **Viral causes:** swine fever, pseudorabies and picorna viruses.

iii. **Fungal causes**: aspegillosis and other fungi.

iv. **Nutritional causes**: iodine, iron, vitamin and calcium deficiencies.

**Other causes** –old age, hormonal (estrogen) and various stresses.

## III. Bacterial Diseases

Bacterial diseases are identified and diagnosed on the basis of clinical manifestations of disease and various diagnostic investigations/tests. The diseases should be diagnosed as early as possible so that the animal may be given proper treatment and its life may be saved. Some of the common bacterial diseases are:

### 1. Erysipelas (Diamond skin disease)

Erysipelas disease in pigs is a contagious bacterial disease, primarily affecting pigs and secondarily sheep and turkey. The characteristic changes include a long term pathological changes in skin, heart and various joints of the body. It mainly affects adult pigs. Mortality is not very common but there is severe adverse impact on growth rate after the disease.

It is caused by *Erysipeothrix insidiosa* (*Rhusiopathy*).

#### *Symptoms*

**In acute form** the temperature rise to 105° F or more, the animal is lethargic, unwilling to rise because of painful joints, anorectic and cyanotic in their extremities. If the animal survives, the skin of the whole body has small, separate pink points. The whole body has areas of inflammation which may appear red, pink or light yellow. After 2-3 days, the classic "diamond" skin lesion develops over the body surface.

**In subacute form:** fever appears as in acute form but does not exist for longer duration. The animal is excited and starts walking. Faeces are also normal.

**In chronic form:** patches appear in the skin of the animals which exists till death. Margins of teats, tails, hooves decompose and fall down. Chronic form may show arthritis, swollen and firm carpuses and hocks, skin necrosis reduced libido in the boar. There may be difficulty in respiration and sitting with hips. Raised skin lesions progress to sloughed areas. In extreme cases, the ears and tail may be sloughed. Vegetative endocarditis is a frequent change.

#### *Diagnosis*

Diagnosis is based on the clinical signs, of which diamond skin lesion are the most characteristic. Isolation of the etiological agent is important for a definitive diagnosis. It may be isolated from skin and joints. Chronic infection is diagnosed by testing the serum through plate agglutination test. Animal inoculation test may be performed.

### *Treatment*

The disease is treated with penicillin and anti erysipelas serum. Usually penicillin is administered mixed with antiserum. If the disease is not very serious, penicillin alone is sufficient. Chronic form of the disease produces permanent pathological changes in joint and the heart because of which the treatment is ineffective at this stage. Corticoids may be used.

### *Control*

The complete eradication of disease is nearly impossible. For control of the disease the affected animal may be slaughtered immediately. Proper quarantine measures should be taken for newly purchased animals before mixing with old stock. During spread of the disease, other animals coming in contact may be may be protected for 1-2 wks by administering 5-10 ml anti- erysipelas serum as per age. Vaccination with either modified live or killed organism is effective in controlling eysepelas in a herd. First Vaccination at 6-10 wks of age and second dose at 2-3 wks later at dose rate of 1ml s/c or 2 ml orally is effective.

### *Public Health Importance*

The disease may occur in human as well. The disease is mostly restricted to people working in packing and keeping of animal products and mostly restricted to hands arms. The affected area become inflamed, painful; have itching and irritation with eruptions on the affected skin. The affected human may have secondary complication exhibited by joint and heart diseases.

## 2. Tuberculosis in Pigs

This is a chronic contagious disease caused by *Mycobacterium tuberculosis.* It is an important zoonotic disease.

### *Clinical Manifestations*

Pigs have symptoms like cattle. The joints and brain are involved. Lesions are mostly confined to lungs, alimentary tract, bones and skin.

### *Diagnosis*

Diagnosis is based on:

1. **History:** the history of the herd or area gives an idea of occurrence of the disease. Many herds have a history of Tuberculosis.
2. **Clinical signs:** proper and close observation of the animal may give rise to suspect the disease. Lymph node examinations and X-rays with ultrasound may help in diagnosting the alment.
3. **Post mortem examination:** on post mortem examinations, the lungs may reveal calcification and casseation.
4. **Bacterial examination**
5. Laboratory animal inoculation
6. Allergic or tuberculin testing

***Treatment***

Treatment is not recommended in affected pigs because it requires much time and money and this is not economically feasible.

**Control**

i. Any person with tuberculosis should be restricted from entering pig houses.
ii. Tuberculosis affected cattle, sheep, goats and poultry should not be mixed with pigs in grazing areas.
iii. Pig house should be regularly and properly cleaned.

**3. Brucellosis**

Also known as Bang's disease or contagious abortion. It is caused by *Brucella suis.* Brucellosis is very important from public health point of view. The main transmission is through contaminated water, feed or vaginal discharges of infected sows. Infected males also transmit the infection.

***Symptoms***

Pigs develop infertility, abortion, joint ill, lameness, debility, birth of weak young ones and scrotal swelling in males etc.

***Treatment***

Normally, treatment is not recommended because medicines have nealy no effect or minimal effect.

***Control***

i. The infected pig should be isolated.
ii. The infected animal should be slaughtered.
iii. The infected male should not be used for breeding purposes.
iv. The aborted foetus and placenta should be covered underground.
v. Quarantine should be done before introducing new animals and checked for the infection.
vi. The animal house should be properly cleaned.
vii. Vaccination is not very effective.

**4. Listeriosis**

It is a contagious disease caused by *Listeria monocytogenes.*

***Symptoms***

There is affection of the central nervous system and consequently associated manifestations develop. There is high rise of temperature (106-107° F), confusion, stretching of the hind limbs develop. Severe mortality in suckling piglets.

***Control***

i. The infected animals may be diagnosed and isolated from the healthy ones.
ii. Rats and other organisms which act as a carrier may be destroyed. Feed and water of animals may be kept away from these.
iii. No vaccination is available.

### *Treatment*

Symptomatic treatment is attempted. Penicillin and sulphonamides have favourable impact on initial stages. Animals which survive have permanent pathological alterations in their brain.

### 5. Haemorrhagic Septicaemia

The disease is also named barbone and shipping fever. It caused by *Pasteurella multocida.*

### *Symptoms*

There is high rise of fever, there is watery discharge from nostrils and mouth. There is swelling in the neck and animal stop eating and drinking. Probability of abortion increases in pregnant sows.

### *Treatment*

Injections of chlortetracycline and oxytetracycline at 6-8 hours interval for 5-7 days may be given.

Dexamethasone and antihistamines like chlorphenaamine maleate injections may also be given simultaneously. Liver extract and vitamin B-complex may be given as supportive therapy.

### 5. Tetanus

Also known as locked jaw, wooden horse disease. It is caused by bacteria named *Clostridium tetani* which can survive for several years in soil and open air and is neither destroyed by acids.

### *Symptoms*

All the muscles of the body become stiff. This is due to the release of neurotoxin by the organism. Jaws are completely locked. Limbs and ears become very hard. The muscles of respiration become stiff and the animal dies.

### *Treatment*

Penicillin is proved to be useful. Muscle relaxants may be advocated, like chlorpromazine. Antitoxin may also be used.

### *Control*

1. Avoid sharp and deep wounds and if such wounds occur, immediately anti tetanus serum or tetvac injection may be given within 24 hour.
2. After any surgery, injection of tetvac may be given.
3. Alum precipitated toxoid may be used.

## IV. Viral Diseases

The main viral diseases are as follows:

### 1. Foot and Mouth Disease (FMD)

Foot and mouth disease is a severe contagious disease of animals. it is caused by an aptho virus of picornaviridae family. It is an acute and very contagious viral disease characterised by fever, formation of vesicles on the coronary band and less frequently, on the lips and tongue. Morbidity is high but mortality is low, except amongst young pigs in which sudden death commonly occurs. The disease is transmitted in pigs from cattle and buffaloes.

### Clinical Signs

The sudden onset of severe lameness is the commonest finding in affected pigs. The infected animals has fever (upto 106 °F). The back may be arched, reluctance to move is common and movement may be accompanied by squealing. vesicles appear as raised white areas 0.5-1 cm in diameter on the dorsum of the tongue and on the snout and may rupture readily to leave small ulcers. Frothy saliva, loss of apptite but the animal has an increased desire to drink water. Ulcers are seen in the mouth and feet. Limbs get lame. The animal often licks the areas around hooves and there is salivation.

### Treatment

The mouth of the infected animal may be washed with 2 per cent alum solution in water or 5 per cent potassium permanganate solution in water. Limbs should be washed with 1 per cent copper sulphate solution. When glycerine is applied with this on the affected ulcers of the tongue, it provides a great relief from pain and sufferings and also helps in wound healing. Supportive therapy like anti – histamines and anti inflammatory drugs may be used. Antibiotics and sulpha drugs are quite helpful. The animal may be provided palatable, easily digestible and energy rich diet.

### Prophylaxis and Control

Vaccination may be used strategically in areas surrounding an outbreak in combination with the existing slaughter policy. Complete care on hygiene and management should be given.

## 2. Swine Fever (Hog Cholera)

It is caused by a virus of the Togaviridae family and pestivirus genus. Ninety percent of the affected pigs die and most susceptible age group is piglets of 2 months age.

### Clinical Signs

Swine fever is a highly contagious disease affecting only pigs. The disease appears in two forms, namely the chronic and acute form. The incubation period in chronic form is 5-10 days.

Initial clinical signs include sudden death in animals, dull and depressed appearance of pigs, loss of appetite, dullness, animal stands still with lowered neck and tail, disinclination to move, conjunctivitis and a fever (40.5 to 41.5 °C). There is a characteristic weaving staggering gait due to weakness in the hind quarters and encephalitis is a characteristic. Other manifestations include an initial constipation followed by diarrhoea and vomition, cyanotic appearance of body, reddening and swelling of eyes and nervous symptoms include walking in circles, staggering and convulsions and muscular twitching.the affected animals die in 5-7 days. Acute form is uncommon. It mostly affects adult animals. The main symptoms include weakening of pigs, special deformities in skin including falling of hairs and cyanosis of body.

### *Treatment*

There is no effective treatment. Antimicrobial therapy may control secondary bacterial infection. Good management practices should be adopted.

### *Prevention*

Vaccination with tissue culture vaccine may be used at 7-9 weeks, the second vaccine at 6-7 months and later at one year interval. piglets from unvaccinated sow may be given at 2 weeks of age and from vaccinated sow at 2 months of age and later at yearly interval at the dose of 1 ml s/c.

## 3. Pseudorabies

Also called Aujeszky's disease. It is caused by herpes virus. It is a contagious viral disease identified by excessive itching and irritation with death in 12-18 hours. The infected pigs are an important source of disease transmission to healthy pigs. Infection spread via nasal discharge, saliva, milk and individual direct contact between pigs.

### *Clinical Signs*

The first and important sign of the disease is excessive itching in the affected animal. The animal continuously licks the affected areas and rubs the body against available tree trunks, walls and other inanimate objects. The hind limb becomes affected and the animal falls on the ground. The temperature of the body rises (105 to 107° F). Following the appearance of first and initial signs of the disease, the affected animal dies in 2-3 days.

The clinical signs of Aujeszky's disease differ according to the age of the animals affected. They are most severe in pigs aged 0-4 weeks in which morbidity and mortality may reach 100 per cent in newborn animals declining to 40-60 per cent in 4 week old animals.

### *Treatment*

No effective treatment is available for this disease. For disease prevention, inactivated vaccine may be resorted to. Use of infected serum may reduce mortality rate in pigs.

### *Control*

Eradication programme must aim at herd eradication which can be done by:

a) Test and remove
b) Vaccination and hysterectomy
c) Vaccination
d) Isolation
e) Antibody free nucleus herds

## 4. Rabies

It is also known as Lyssa or hydrophobia. It is caused by a virus and is an important viral disease of man and other mammals, which is highly contagious and fatal with symptoms of excitement, paralysis of muscles, salivation, biting of animals and man, madness and death.

The disease is mainly transmitted by the bite of infected animals. The main transmitters are dogs, fox, jackals and cats. The virus is present in saliva of affected animals. Rabied dog/animal bite other animals and the animal gains entry into the body through neurons. The forms of the disease are visible.

One is the dumb form and the other is the furious form. in dumb form, the animal is quite, there is salivation from the mouth, the jaw hangs down and there is a change in voice.

In furious form, the animal becomes excited and attempts to bite objects around it. It may bite and attack the owner , head is raised and stiff, dwells with the legs and hits at the wall.painfull sound comes from throat. There is salivation and froathing from the mouth. Because of difficulty in chewing and swallowing, the animal becomes disinclined to eat. Initially, there is constipation which is later followed by diarrhoea. Slowly the hind portion of the body becomes useless. Death occurs in 3-6 days. In dumb form of rabies, the animal does not eat. There is paralysis of the hindquarters. Salivation occurs from mouth and a painfull sound comes and the animal dies within a week.

### *Control*

I. Isolate the affected animal from other healthy ones. On diagnosis of the disease the affected animal may be slaughtered so that the disease does not spread to other animals. There should be proper disposal of carcass of the sick animal.

II. The biting wound by infected animal should be immediately washed with antiseptic solution and potassium permanganate solution may be applied. The wound should be repeated solution may be applied. The wound should be repeatedly washed with soap and water and later washed under running water.

III. Street dogs and wild animals may be kept at a distance and should be avoided from coming close.

When infected animal bites a pig, then post bite anti-rabies vaccine should be given to the bitten pig on 0,3, 7,14,30 and 90 days post bite.

## 5. Transmissible Gastroenteritis (TGE)

It is caused by a corona virus.

### *Clinical Signs*

Initially, the disease is mostly observed in growing and finishing pigs with some vomiting and diarrhoea. Afterwards, the disease appears in a severe form in the farrowing piglets affecting most of the animals within one to two days. The clinical manifestations include rough haircoat, shivering, vomiting, refusal to nurse, and a severe thirst. Followed by a severe watery very putrid and characteristically odored diarrhoea appears leading to severe dehydration and weakness. The animal dies within 48 to 120 hours. In acute form, the mortality may be 100 per cent in pigs under 7 days old. Lactating sows may also become infected. They exhibit vomition, diarrhoea and may stop lactating. In weaned pigs, the morbidity may be 100 per cent with nearly zero mortality.

The persistent form of the disease is called chronic or enzootic TGE. The carrier pigs transmit the infection. It is more common in herds with multiple or continuous farrowing which makes the available pigs susceptible at all times. Ten days old to weaning appears to be the most common age of infection.

### *Treatment*

It is recommended that all litters over one week of age be weaned and fed either gruel or a proper milk replacer. Fluid therapy may be adopted. Antibiotics with vitamin B- complex may also be used.

### *Control*

I. Isolate swine from dogs and foxes.

II. Do not introduce new pigs during the farrowing season.

III. New pigs should be quarantined for two months.

IV. Modified live virus vaccines are available.

## 6. Rotavirus Diarrhoea

It is caused by rotavirus. It is an important cause of diarrhoea in young swine.

### *Clinical Manifestations*

Incubation period is 2-4 days. It is common in 3-6 weeks old animals,. There is anorexia, depression, diarrhoea, reluctance to move. Adult animals generally do not show any clinical signs of infection, although has been observed in gilts. Diarrhoea may be initially watery, then creamy, followed by profuse yellow to green liquid faeces. There is rapid loss of condition. Rotavirus diarrhoea may reccur as recovery from one type does not prevent infection with others. Dehydration is more evident in younger piglets, and in those with prolonged diarrhoeic periods. More than 80 per cent animals may be affected. Mortality may reach 20 per cent in infected animals.

### *Treatment*

Fluid therapy, antibiotics and vitamins as indicated for TGE are recommended. Withholding milk from affected piglets. Water should always be available.

### *Prevention*

Proper cleanliness and hygiene. Prevention may be achieved by stimulating maternal immunity. Vaccines are available commercially for the immunization of pregnant gilts, sows and nursing pigs.

## 7. Swine Flu/Swine Influenza

Swine influenza also called pig influenza, swine flu, hog flu and pig flu, is an infection by any one of the types of swine influenza virus.this disease occurs in pigs of all ages. Mortality rate is low but there is decrease in growth rate after the infection.

It is caused by type A influenza virus infection.

Currently, 16 haemagglutinin (HA) subtypes and 9 neuraminidase (NA) subtypes are identified. Three main subtypes currently circulating in the pig population are classical swine influenza virus and reassortant viruses of H1N1, H3N2 and H1N2.

### *Signs and Symptoms*

#### *In Swine*

In pigs, influenza infection produces fever, lethargy, sneezing, coughing, difficulty breathing and decreased appetite. In some cases the infection can cause abortion. Although mortality is usually low, the virus can produce weight loss and poor growth, causing economic loss to farmers.

#### *In Human*

Symptoms of swine flu are similar to most influenza infections- fever, cough, nasal secretions, fatigue, and headache, some patients also get nausea, vomiting and diarrhoea.

### *Treatment and Control*

There is no effective treatment, although antimicrobials may reduce secondary bacterial infections. Expectorants may help relieve signs in severely affected herds.vaccination and strict import controls are the only specific preventive measures.

## V. Vaccination Schedule in Swine

### Mycoplasmal Disease

#### Enzootic pneumonia

The disease is caused by *Mycoplasma hyopnemonie.* It basically affects the respiratory tract. It is mostly seen during winter. The mortality rate is high.the growth rate is decreased upto 13 per cent.

### *Symptoms*

The disease occurs in two forms:

I. Acute form: pigs of all age group are affected and the mortality rate is nearly 100 per cent. There is difficulty in respiration. Fever may not be present.it may be confused with swine influenza.

II. Chronic form: it is the most common form. Young ones 1-2 months of age are more affected. There is coughing, growth rate decreases and there is no fever in this form.

### *Treatment*

There is no satisfactory treatment. Only the severity of the infection may be reduced: tylosin tartarate @ 5 mg/kg bwt may be given for 10 days.

### *Control*

i. The infected group of animals may be kept isolate from the healthy group of animals.
ii. The infected stock may be replaced with a healthy stock at the earliest.
iii. Proper arrangement should be made to keep the pig houses hygienic and clean.

### Mycoplasmal Arthritis

It is a disease of young pigs (3- 10 weeks of age) the disease is caused by two species of mycoplasma. The fist one is *Mycoplasma hyohernis* and the other one is *Mycoplasma hyosinova*.initially it affects the respiratory tract and later produces disease in the joints. Mycoplasma hyosinova affect tonsils.

#### *Symptoms*

In infection with *Mycoplasma hyohrnis*, there is fever, loss of appetite and difficulty in respiration with the animal remaining constantly dull and depressed. there is swelling of the joits and the animal becomes lame.

In infection with *Mycoplasma hyosinova* the disease is quite severe. Lameness is very severe and may affect one or several limbs. Many joints may be affected.

#### *Treatment*

Tylosin @ 10-20 mg/kg body weight or alincomycin @ 2.5 mg/kg body weight may be given as injections with betamethasone.

#### *Control*

Tylosin and tetracycline in drinking water or feed may be preventive for this disease.

## VI. Parasitic Diseases

These are common in pigs. These do not allow the pig to grow properly and also reduce their market value. Animals should be routinely and timely drenched with anthelmintics.

### A. Endoparasites

Helminthic infestations are of three types:

1. Trematodes
2. Cestodes
3. Round worms

Pigs are mostly infected with fasciolopsis. Fasciolopsis is found in the intestine of pigs. The affected pigs exhibit diarrhoea, stomachache, and restlessness. The pig get infected by eating it at the metacercaria stage of its life cycle which comes through grass and leaves. For diagnosis, faecal examination of pigs should be regularly carried out and affected pigs should be treated with anthelmintics like albendazole, praziquintal and niclosamide etc.

### Fasciola Hepatica

It mainly affects the liver. Snails are the intermediate hosts. Metacercariae encysted on plants that have been under water transmit the infection when they are ingested by pigs.

### Tape Worms

#### *Taenia solium (Cysticercus cellulosae)*

It lives in the skeletal muscle. The human is the primary host. Swine and humans may be intermediate hosts. Its incidence is related to sanitation.

### *Clinical Manifestations*

There are no characteristic clinical manifestations in affected pigs but by eating the mass of infected pigs, the human pick up the infection of the tapeworm, *Taenia solium.*

## Round Worms

Many round worms are found in pigs like *Ascaris suum, Oesophagostomum, strongyloides ransomi* and *Globocephalus cacanthrax.*

### *Ascaris suum*

Ascariasis is a worldwide problem mainly involving pigs. This roundworm is found in the small intestine of pigs.it is the largest gastrointestinal parasite of swine. The adults are found in the small intestine and the larvae may be found in the liver and lungs.

### *Clinical Manifestations*

Cough is the main sign. It causes restlessness, diarrhoea or constipation, pneumonia, mucoid cough and retarded growth in affected pigs. If present in more numbers, it may completely obstruct the intestines.the affected pigs show poor growth rate and may even die in severe infection.

### *Oesophagostomum dentatum*

*Oesophagostomum spp.* Infect the caecum and large intestine. Adult worm is found in the large intestine and its larval stage is present in various layers of the intestines. It is ubiquitous and probably the most common parasite of adult swine.

### *Clinical Manifestations*

Its infection causes watery diarrhoea, retarded growth and loss of appetite in affected pigs. The larvae form nodules in the intestine of pigs by which there is disturbance in the intestinal absorbtion of digested material causing weakness and retarded growth in affected pigs.

Severe signs of anorexia, constipation, severe diarrhoea and emaciation may be in conjunction with secondary pathogens.

### *Stephanurus dentatus*

It is found in the urinary tract. Stephanurus, the kidney worm, is a strongylid nematode. Its only natural host appears to be swine. It is a round worm found in kidneys of pigs, larval stages of which are found in the liver and other organs.

It is found in perirenal fat, ureters, kidney and liver. Because of its larval migratory phase, ectopic sites are many like pancrease, lumbar muscles, spinal cord and lungs.

The affected pigs exhibit retarded growth, loss of appetite and weakness.

### *Trichinella spiralis*

It is found in the small intestine.Adults are found in the small intestine and larvae in the blood and lymphatics and eventually become encysted in the diaphragm and messeter, lingual and intercostals muscles but may be in any striated muscles.

It is less pathogenic for swine than for humans. It causes decreased weight gains and intense muscular pain. Most of the infected pigs recovered and made rapid weight gain.

### *Trichuris suis*

It is found in the large intestine. *T. suis* is a nematode that infects, pigs, wild boar, monkeys and humans. Whipworm are found in caecum, colon, and appendix (human), with migration limited to the walls of the gut.

### *Clinical Manifestations*

Includes anorexia, diarrhoea with mucus and blood in the faeces, anaemia,labored respiration, incoordination, emaciation, retardation of growth, dehydration and death.

## B. Protozoal Infections

### a. Babesiosis

It is also known as texas fever, red water fever and cattle tick fever. The disease in pigs is caused by *Babesia trautmanii* and *Babesia paranisitie.* The disease is transmitted by the tick boophilus which transmit the infection while biting. The protozoa are present in the blood of the host. When the boophilus sucks blood from an infected animal, the infection is transmitted and the healthy animal acquires up the infection. The clinical manifestations include high fever, loss of appetite and weakness. There is an increase in the respiration and pulse rates. Later ther is development of jaundice and the urine becomes red or brown. Pregnant sows may abort.

### *Treatment*

Diminazine aceturate and imidocarb are drugs of choice. With this, to recuperate from anaemia and blood loss, intravenous blood transfusion by a veterinarian only and as supportive therapy vitamin B complex, liver extract, antihistaminics and fluid therapy may also be advocated.

### *Control*

i. Ticks may be controlled by spray with ectoparasiticidal drugs in pigs houses.
ii. Pigs should be routinely washed with malathion or cypermethrin solution.
iii. Diseased pigs should be isolated from the healthy ones
iv. Use of drugs like imidocarb greatly reduces the chances of occurrence of infection.

### b. Coccidiosis

The disease is found in all domestic animals and is mostly confined to young ones. There is diarrhoea and dysentery. The new borns are mostly affected 5-15 days. Ingestion of contaminated feed and water spreads the infection.

In pigs, the disease is caused by the protozoa *Isospora suis, Eimeria debliecki, Eimeria scabra, Eimeria cerdonis and Eimeria perminuta.*

These are intracellular parasites of the intestinal tract.the disease is called milk scour, white diarrhoea or three week diarrhoea.

In newly born piglets, the symptoms appear at 5-15 days of age. The animal becomes dull and depressed and stops taking milk. In later stages, severity and frequency of diarrhoea becomes quite high. This complication leads to excessive loss of water and ultimately causes dehydration which needs immediate therapy.

***Treatment***

Sulphadimidine, nitrofuran and amprolium salt are the medicines used in this disease.

***Prevention***

1. The bedding used in piggery should be routinely changed and cleaned.
2. The pig houses should be regularly and properly cleaned and kept hygienic.
3. There should be rotational grazing in pastures.
4. Chlortetracycline and sulphonamide should be given mixed in feed to pigs.

To protect pigs against endoparasites the pig owner should adopt the following practices:

i. Pigs should not be allowed to graze in lands with wetness and snails.
ii. Faecal examination for parasites should be done regularly and the moment any infection is detected, it should be treated.
iii. The pig farm should be kept neat and tidy.
iv. Infected animals should be kept separate from healthy animals and their faeces should not be allowed to accumulate in the animals sheds.
v. The surroundings of the pig houses should be kept clean and disease transmitting ticks and insects should also be prohibited from entering and multiplying.
vi. The young piglets should be kept isolated from the adult ones.
vii. The pigs should be regularly dewormed with anthelmintics.

Some of the anthelmintics used are:

**a. Pipearzine salts**

Piperazine may be given in the feed or water @275-440 mg/kg body weight. Piperazine citrate for one day in feed. Piperazine hexahydrate is given in drinking water.

**b. Hygromycin B**

It is effective against mature stages of *Ascaris* and *oesophagostomum.* It can be given in the feed for consumption over several weeks (for 6 weeks before farrowing to sows and during nursing and to weaned pigs until 6 m of age. It is mixed with the feed @ 12g/ton.

**c. Organophosphate Compounds**

Diclorvos can be given in the feed @ 11.2 – 21.6 mg/kg body weight.

**d. Imidathiazoles**

Levamisole acts against gastrointestinal, respiratory and urinary tract nematodes. Levamisole hydrochlroride @ 8mg/kg body weight is given in feed or water.

**e. Benzimidazole Carbamates**

Thiabendazole @ 50mg/kg bwt can be used.mebendazole, cambendazole, albendazole (albomar), fenbendazole (fentas plus) and oxfendazole are other drugs.

## C. Ectoparasites

Various ectoparasites are found in the body of pigs. They themselves harm the pigs and also act as transmitters of various diseases in pigs, thus additively having an adverse effect on the health of the pig and giving an economic loss to the pig farmers.

Ectoparasites may cause unthriftiness and decreased weight gains. This will lead to more feed required per unit gain in body weight than that required by a parasite free pig. Ectoparasites include mites and ticks of the class Arachnida. Whereas lice, fleas and flies belong to the class insecta.

Some ectoparasites create wounds by biting the pigs in which flies like the blow fly and screw worm fly lay eggs and the larvae which hatch from them enter deep into the wounds and continuously harm the host. Some ectoparasites transmit various bacterial, rickettsial, viral and protozoal diseases from one pig to the other. Some ectoparasites like bugs and flies bite on the skin of the host which makes them restless. The main ectoparasites of pigs include ticks, mites, lice, bugs, fleas and flies which may be controlled as follows:

| *Sl.No.* | *Ectoparasite* | *Medicines* | *Usage* |
|---|---|---|---|
| 1. | Ticks | i. Toxophan | 0.5 per cent spray |
| | | ii. Chlorfenvinphos | 0.5 per cent spray |
| | | iii. Coumaphos | 0.5 per cent spray |
| | | iv. Diazinon | 0.5 per cent spray |
| | | v. Deltamelthrin | 0.2 per cent spray |
| | | vi. Dioxathiozone | 0.01 per cent spray |
| 2. | Mites<br>i. Sarcoptic mites | a. Ivermectin | 0.3mg per kilogram body weight asan injection (subcutaneous), repeated after 28 days. |
| | | b. Carbaryl | 1 per cent spray (asuntol 0.05 per cent spray) |
| | | c. Deltamethrin | 0.02 per cent spray |
| | | d. Benzyl benzoate | 25 per cent as local application on skin |
| | ii. Demodex (Follicular mites) | Ivermectin | As above |
| 3. | Lice | Fenthion | 3 per cent spray |
| 4. | Bugs | Malathion | 6 per cent spray |
| | | Lindane | 1 per cent spray |
| | | Kilex carbaryl | 5 per cent spray |
| 5. | Flies | Malathion | 2.5 per cent spray |

# 6

# Case Studies

## Case Study I: Comparative Study of Traditional Farming System and Improved Farming System (Ansari *et al.*, 2013)

The comparative study of traditional farming system and improved farming system was carried out in another survey in the Chandel (Chandel Khulen and nearby village), Churachandpur (Vazing and village nearby village) and Ukhrul (Ramva village and nearby village) hill districts of Manipur during 2012–13. A total of 70 farmers/districts, with a cluster of thirty five farmers for improved farming systems (IIFS) and thirty five for conventional farming system (CFS) were surveyed for this study. The average holding of each farmer was 0.64 ha in Chandel (Chandel Khulen), 0.65 ha in Churachandpur (Vazing village) and 0.87 ha in Ukhrul (Ramva village). Seven components, *i.e.* paddy, vegetables, fruits, piggery, backyard poultry, fishery and water management were studied for comparisons based on the criteria as mentioned in following table. During the study period, the multi disciplinary scientific personnel from ICAR Manipur Centre have visited the farmer's household on pre-scheduled time and collected necessary information from interviews using structured questionnaires. The interviews were conducted individually as well as in groups with the respondents. However, while interviewing in the groups, participant observation method was also employed to observe the farmer's perceptions and recording knowledge on profitability of different farming systems. Several visits in the rural areas were conducted to collect the information and verification of these practices. The data on cost of different inputs and return from outputs for different systems (CFS and IIFS) were also recorded for each farmer. The data collected were analyzed through the use of descriptive statistics using means and standard error for explaining the outcomes of findings.

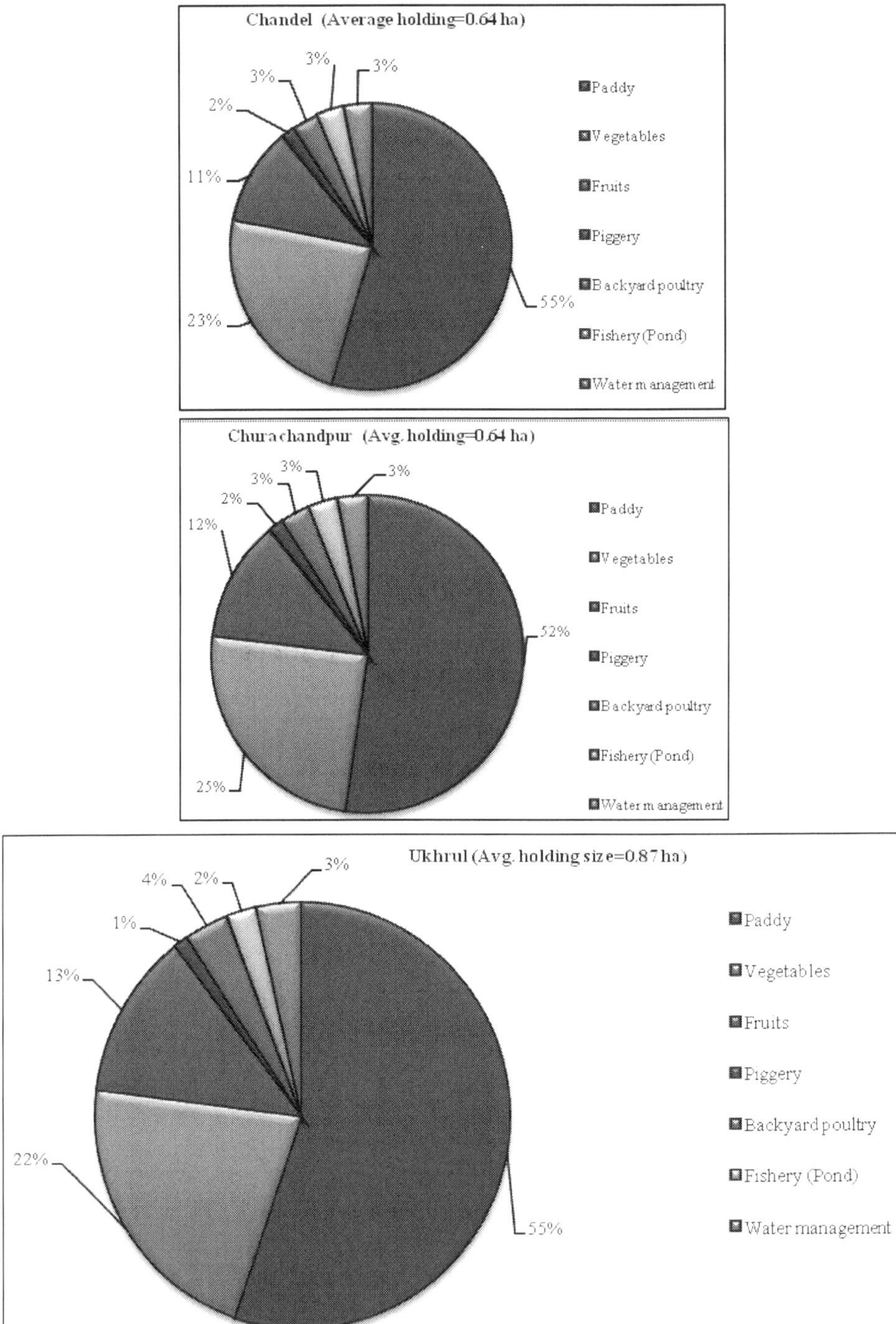


**Area Under Different Components in Integrated Farming System in Chandel, Churachandpur and Ukhrul (The average holding size of farm was 0.64, 0.65 and 0.87 ha).**

### Price of different Enterprises and Average Labour Requirement

| Sl.No. | Particulars | Price (Rs) | Enterprises | Labour Engaged (Days) | |
|---|---|---|---|---|---|
| | ***Main Product*** | | | ***CFS*** | ***IIFS*** |
| 1. | Paddy | 10/kg | Paddy | 20 | 30 |
| 2. | Vegetables | 10/kg | Vegetables | 18 | 25 |
| 3. | Fruits | 20/kg | Fruits | 15 | 23 |
| 4. | Pork | 200/kg | Piggery | 50 | 70 |
| 5. | Piglets | 2000/piglet | Poultry | 70 | 90 |
| 6. | Chicken | 100/kg | Fishery | 30 | 39 |
| 7. | Egg | 8/egg | Water management | 8 | 22 |
| 8. | Fish | 100/kg | Total | 211 | 299 |
| 9. | Water | 10 paise/liter | | | |
| | ***By product*** | | | | |
| 10. | Paddy straw | 1/kg | | | |
| 11. | Poultry manure | 6/kg | | | |
| 12. | Pig manure | 6/kg | | | |

### Criteria Followed for the Selection of the Farming Systems

| Sl.No. | Components | Criteria | CFS | IIFS |
|---|---|---|---|---|
| 1. | Paddy cultivation | a. Variety<br>b. Seed quality<br>c. Techniques<br>d. Weed management | a. Local or improved cultivars<br>b. Old cultivars<br>c. Conventional package and practices<br>d. Hand weeding | a. Improved varieties<br>b. Good in quality<br>c. Improved package and practices (Some SRI and ICM)<br>d. Cono/rotatory weeder |
| 2. | Vegetable cultivation | a. Cultivars<br>b. Seed<br>c. Method of cultivation<br>d. Manure and fertilizers | a. Local cultivars<br>b. Uncertified seed<br>c. Conventional methods<br>d. Less than recom-mended dose | a. Improved cultivars<br>b. Certified seed<br>c. Improved methods<br>d. Balance nutrition |
| 3. | Fruits cultivation | a. Parental stock<br>b. Training and pruning<br>c. Fertilization<br>d. Irrigation | a. Old and less production<br>b. Not timely and regularly<br>c. Below recomme-nded dose<br>d. Not properly | a. Improved and more production<br>b. Timely and regularly<br>c. Optimum dose<br>d. Proper at certain interval |
| 4. | Piggery | a. Breed<br>b. Feed<br>c. Health management | a. Local<br>b. Based on house hold wastage<br>c. Rarely | a. Cross breed and improved<br>b. Quality feed<br>c. Proper deworming and vaccination |

*Contd...*

**Contd...**

| Sl.No. | Components | Criteria | CFS | IIFS |
|---|---|---|---|---|
| 5. | Backyard poultry | a. Breed<br>b. Feed<br>c. Health management | a. Local<br>b. Scavenging<br>c. Rarely | a. Improved birds<br>b. Quality and concentrated feed<br>c. Proper vaccination |
| 6. | Fishery | a. Fingerlings<br>b. Pond liming<br>c. Stocking density<br>d. Feed<br>e. Fertilization | a. Local<br>b. Rarely<br>c. Below stocking density<br>d. Natural feed<br>e. Raw cow dung | a. Improved<br>b. Proper at certain interval<br>c. Optimum stocking density<br>d. Natural + concentrated<br>e. Fertilization with well rotted organic and inorganic fertilizers |
| 7. | Water management | a. Water harvesting unit | a. Lacking | a. Jalkund/Farm pond |

**Crop Varieties and Animal Breeds under different Enterprises**

| Sl.No. | Components | CFS | IIFS |
|---|---|---|---|
| 1. | Paddy cultivation | Drumphou, Priya, Jatra, Ayangleima | RCM-5, RCM,9, RCM-10 and other improved cultivars |
| 2. | Vegetable cultivation | Local varieties | Cabbage: Rareball, Wonderball, Green hero<br>Cauliflower: Early Hemlata, White flush, Shweta<br>Broccoli: Green magic, Aishwarya, Puspa<br>Tomato: Nidhi, Swaraksha, Avinash-2 etc.<br>Okra: Arka anamika, Prabhani Kranti, Pusa Sawni |
| 3. | Fruits cultivation | Local cultivars | Banana: Grand nine, Jaint Cavendis, Robusta<br>Orange: Khasi mandarinKachai lemon<br>Passion fruit: Kaveri<br>Pineaaple: Kew, Queen<br>Papaya: Taiwan, Ranchi, Coorg Honey Dew |
| 4. | Piggery | Local Breed | Cross breeds |
| 5. | Backyard poultry | Local birds | Gramapriya, Vanraja, Improved |
| 6. | Fishery | Catla, rohu | Catla, rohu, mrigal, grass carp, common carp etc |
| 7. | Water management | Lacking | Construction of Jalkund/Farm pond |

# Salient Observations

## Paddy Production

Rice is the staple food for more than 90 per cent population of Manipur and is the major component of all the farming system of Manipur. Among the hill districts surveyed, Ukhrul has recorded highest paddy production. The production of paddy in CFS was much less in the present study as compared to rice grown under IIFS. On an average, 5087.1 kg per hectare rice yield was observed under IIFS as compared to

**Production of Paddy (A), Vegetables (B) Fruit (C) Pork (D), Piglets (E), Chicken (F), Egg (G) Fish (H) and Rainwater Harvesting (I) in Chandel, Churachandpur and Ukhrul under CFS and IIFS.**

**Standard error bars represent the standard error value and significant level at P=0.05.**

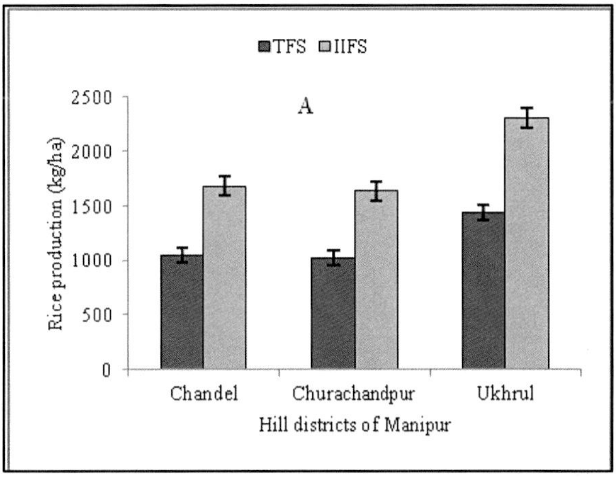


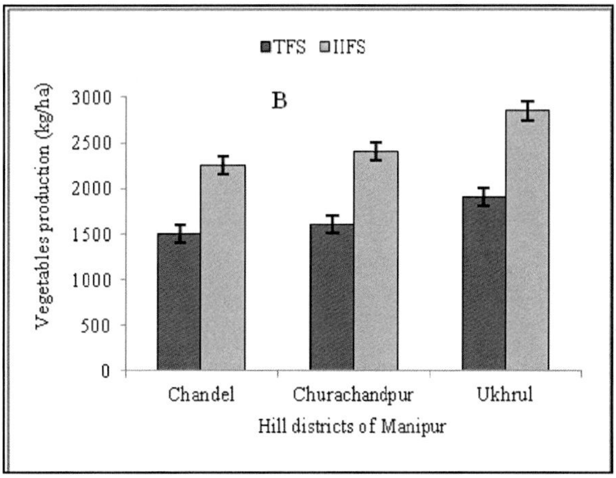


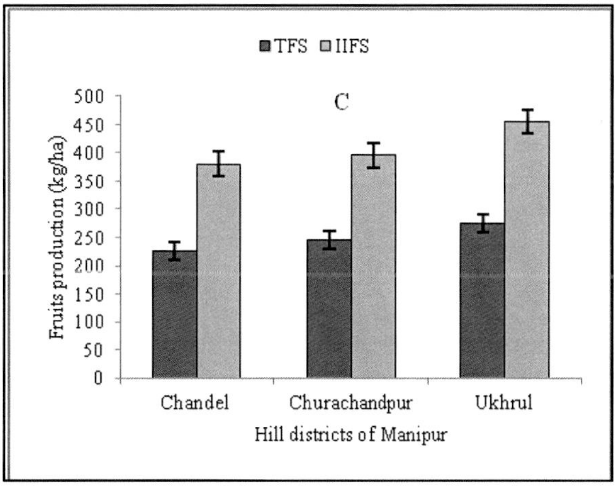


*Contd...*

***Contd...***

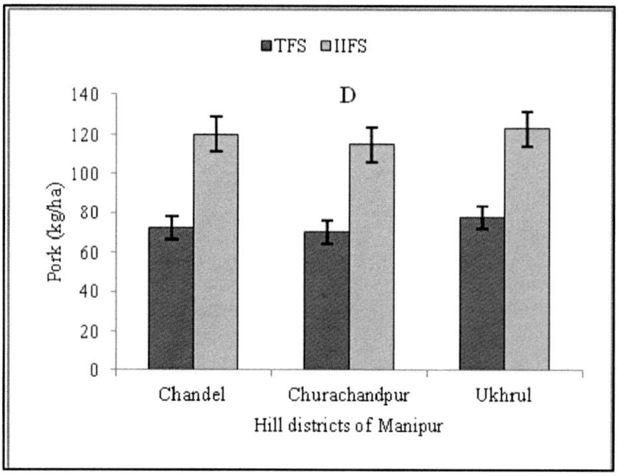
TFS
IIFS
D
Pork (kg/ha)
0
20
40
60
80
100
120
140
Chandel
Churachandpur
Ukhrul
Hill districts of Manipur

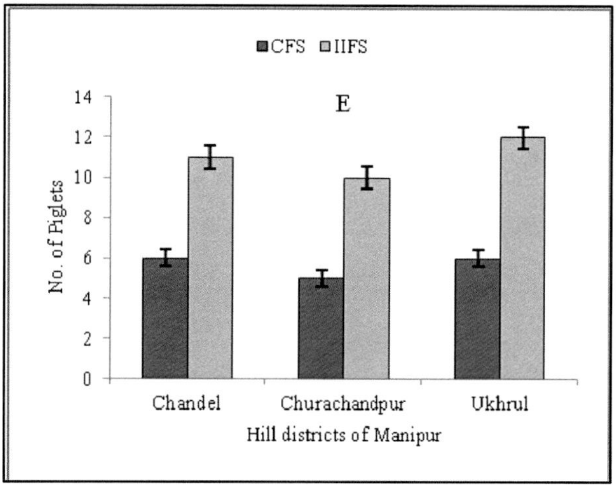
CFS
IIFS
E
No. of Piglets
0
2
4
6
8
10
12
14
Chandel
Churachandpur
Ukhrul
Hill districts of Manipur

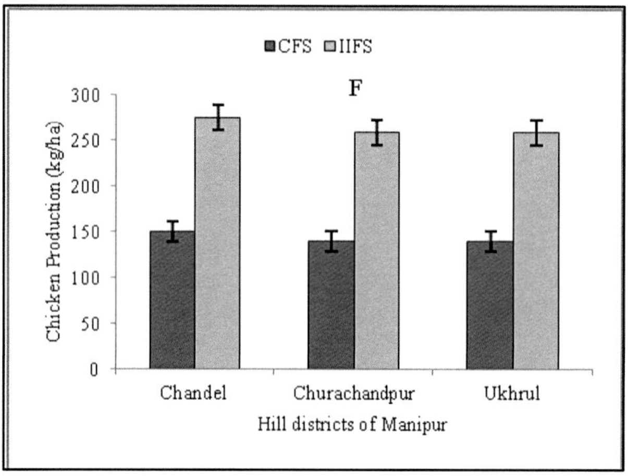
CFS
IIFS
F
Chicken Production (kg/ha)
0
50
100
150
200
250
300
Chandel
Churachandpur
Ukhrul
Hill districts of Manipur

*Contd...*

**Contd...**

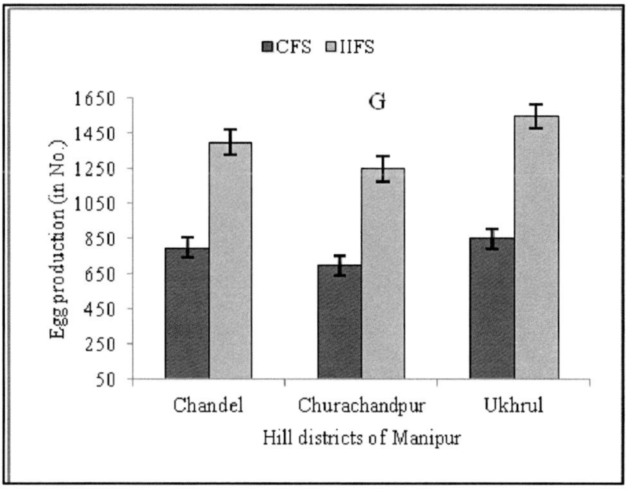
CFS IIFS
G
1650
1450
1250
1050
850
650
450
250
50
Egg production (in No.)
Chandel
Churachandpur
Ukhrul
Hill districts of Manipur

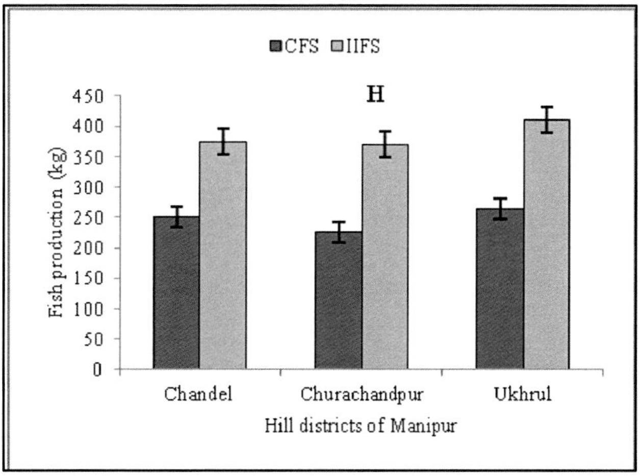
CFS IIFS
H
450
400
350
300
250
200
150
100
50
0
Fish production (kg)
Chandel
Churachandpur
Ukhrul
Hill districts of Manipur

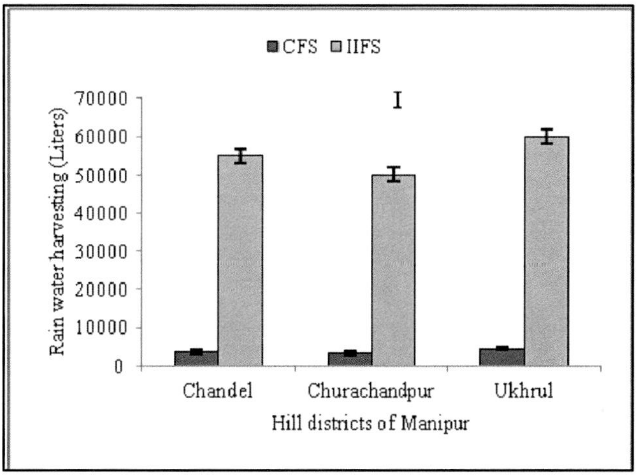
CFS IIFS
I
70000
60000
50000
40000
30000
20000
10000
0
Rain water harvesting (Liters)
Chandel
Churachandpur
Ukhrul
Hill districts of Manipur

3000.15 kg in conventional farming system. Rice grown under IIFS produced 69.5 per cent more yield as compared to CFS. The lower yield of rice under CFS was mainly due to use of poor quality seeds, that they are procuring from neighbouring farmers and use of conventional methods in rice cultivation with very poor nutrient and weed management. Farmers who practice improved farming with certified seeds of improved varieties and scientific practices like system of rice intensification (SRI) with proper nutrient and weed management recorded higher rice yield. Our results are close agreement with Yadav *et al.* (2013), they also studied at Tripura on CFS and IIFS and they found that SRI is the one of the best techniques for paddy cultivation, which significantly increased the paddy yield as compared to conventional system.

## Vegetable Production

Vegetable are mostly grown for household consumption in kitchen gardening and also for commercial cultivation for selling in local market at remunerative price. In Manipur, farmers are growing mainly leafy, root, cruciferous, cucurbits and egg plants at commercial scale. High value vegetables like broccoli, brussels, carrot are also being grown by some farmers. Among the surveyed districts, Ukhrul had higher vegetables productivity followed by Chandel and Churachandpur. In the present study, on an average the production of vegetables in IIFS was 50 per cent higher over CFS. The higher vegetable production in improved farming system might be due to use of quality seeds and better management practices followed by the farmers. The increase in yield of vegetables with improved practices could be attributed to improved vegetative growth, better availability of nutrients at vital growth period and greater synthesis of carbohydrates and their translocation to the storage organs (Yadav *et al.*, 2013). The observations of the present study are in line with the findings of Brahma *et al.*, 2010 and Patel and Rajput 2003.

## Fruit Production

The horticultural crops are grown as integral enterprise of farming system from kitchen gardening to commercial scale. In Manipur, mainly banana, orange (khasi mendarin), kachaai lemon, passion fruits, pineaaple and papaya as being grown. In kitchen gardening (farming system at small scale) fruits production is enough to supply the household requirements. In the present study, Ukhrul has better performing districts in terms of fruit production as compared to other hill districts. On an average, the production of fruit in IIFS was found to be 65 per cent higher as compared to production recorded under CFS. The higher fruit production in improved integrated farming system might be due to use of quality and disease free planting materials and better management practices followed by the farmers. The increase in fruit yield with improved practices could be attributed to improved vegetative growth, better availability of nutrients at vital growth period and greater synthesis of carbohydrates and their translocation to the storage organs (Yadav *et al.*, 2013). Ghosh (2008) also recorded better fruit crops production under improved practices.

## Pig Rearing and Pork Production

Pig farming is one of the choicest enterprises in farming system, where pork is the most preferred meat among tribal farmers. Scientific piggery is a profitable venture because of superior feed conversion ratio, faster growth rate, short generation interval and high proliferation in pigs (Yadav *et al.*, 2013). The indigenous breed of pigs in Manipur is smaller in size with poor feed conversion abilities and average adult body weight is about 45 kg and litter size is 3.5 to 5. The up-gradation of local pigs through cross breeding with the exotic pigs (Hampshire) has given a new dimension to the piggery. The improved breeds are healthy and their body weight is more than the indigenous pigs. In the present study, approximately 63 per cent more weight was found in cross breeds as compared to local breeds. Lekule and Kyvsgaard (2003) also reported poor feed conversion, high mortality rates, low productive rates and poor final products in pigs in conventional method of farming.

## Backyard Poultry

Poultry is playing an important role for human nutrition, family income, employment and income generation in homestead. Breed, strain, reproductive efficiency, and proper management system with regard to housing, balanced feeding, proper nursing, vaccination etc are the major factors in the profitability of poultry farming (Yadav *et al.*, 2013). Poultry farming is the second most important enterprise after piggery in Manipur under farming system. It is more popular in both hills and valley areas. Almost all households have backyard poultry. Grampriya and Vanraja are being reared by the farmers in Manipur. In the present study, 85 and 79 per cent more meat and egg production was recorded under IIFS as compared to CFS. These findings are in line with those of Ershad (2005).

## Fish Production

Fisheries play an important role in the economy of North East, as fish constitutes one of the vital items in the diet of 95 per cent of its population (Dey and Ghoshal, 2011). They are fond of fishes, and cooked for consumption in raw as well as dry form. In the present study, after vegetable production, fishery is the second most profitable enterprise in terms of higher net returns and benefit cost ratio. On an average, the fish production under IIFS was approximately 57 per cent higher than CFS. Our results are in accordance with Yadav *et al.* (2013). Additional production of fish in IIFS might be due to the adoption of scientific fish culture practices. Probably the cumulative effect of maintaining proper ratios of fingerlings with area and different layers of water bodies and uniform feed supply accelerated better fish growth, resulting in increased fish production.

## Water Management

The major water source for cultivating agricultural crops in Manipur is rain water. The rainfall of Manipur varies from 1400 to 2000 mm. In Kharif season, rain water is sufficient for agricultural crops. But, in Rabi season, rain fall is scanty. The farmers having farm pond, Jalkund etc. on their farm can give life saving irrigation for various vegetables and pulses. Present study shows that approximately 3000 to 5000 liters of rain water was conserved as compared to 55000 liters in IIFS. There is 1200 per cent more water conservation under IIFS than CFS in Jalkund, farm pond etc.

## Economics

Profitability in terms of economic returns of any farming system is the integral component for its adoption by farmers. The current study indicates higher variable cost, gross returns, net returns and benefit cost (B:C) ratio in all the components of IIFS as compared to CFS. Among the districts, Ukhrul has better adoption of farming system either CFS or IIFS than other two hill districts of Manipur. The higher variable cost in all the components of IIFS was mainly due to use of quality seed material of improved varieties/breeds, optimum fertilization, proper weed management and pest management with better package and practices. The higher gross returns, net returns and B: C ratios were due to more production of economic products in respective components of IIFS as compared to CFS. The lower cost of production, gross returns, net returns and B:C ratios in CFS might be due to use of poor quality seed material of local varieties/breeds with very poor management practices just for subsistence farming for sustaining the family livelihood. Among the components, vegetable production showed higher variable cost and higher gross and net returns, followed by backyard poultry, fishery and piggery in both the farming systems. Overall, the IIFS showed more expenditure towards variable cost and higher gross and net returns when compared to CFS. The net return is the main parameter for deciding the adoptability of a farming system. The present study showed that the IIFS is more profitable and sustainable as compared to farmer's conventional farming system. Similar results were also reported by Kumar *et al.* (2011); Alam *et al.* (2009) and Yadav *et al.* (2013).

## Employment Generation

A special feature of farming system is value addition due to employment of family members and non farming families. One marginal farmer absorb 593 person days, out of which family labour absorption was 470 days person (MSSRF, 2009). On an average 299 man days were utilised under IIFS as compared to 211 man days under CFS. Moreover, women participation was more (45.5 per cent in IIFS, 24.9 per cent in CFS) as compared to men (36 per cent in IIFS, 33 per cent in CFS). Backyard poultry followed by pig rearing were the top most enterprises for engaging the household men and women for longer duration. Thus, integrated farming system is suitable for ensuring sustainable livelihood for North Eastern people.

There is a tremendous potential for developing farms to their optimum levels by integrating different enterprises in a farming system mode to ensure make agriculture a profitable venture for farmers. It can be concluded that improved farming can play a significant role in increasing production, remunerative returns, and nutrition requirements as well as employment opportunities of tribal populations. The overall findings in terms of production, economic returns and employment generation with seven components of conventional and improved integrated farming system in the present study in Chandel, Churachandpur and Ukhrul hill districts of Manipur revealed that improved farming is better over conventional farming.

**Variable Costs of different Component of Farming Systems at Chandel, Churachandpur and Ukhrul under CFS and IIFS**

| S.N. | Components | Variable Cost (Rs) | | | | | |
|---|---|---|---|---|---|---|---|
| | | Chandel | | Churachandpur | | Ukhrul | |
| | | CFS | IIFS | CFS | IIFS | CFS | IIFS |
| 1. | Paddy cultivation | 4,442.5±86.9 | 6,152.5±81.8 | 4,310.3±124.2 | 6,052.2±105.6 | 6,020.7±172.2 | 8,454.0±141.7 |
| 2. | Vegetable cultivation | 2,642.5±57.8 | 3,520.8±61.4 | 2,811.3±118.5 | 3,608.3±108.3 | 3,331.2±127.5 | 4,500.8±112.7 |
| 3. | Fruits cultivation | 1,501.7±30.3 | 2,241.7±79.5 | 1,543.5±57.0 | 2,315.0±98.3 | 1,543.3±106.5 | 2,315.0±114.3 |
| 4. | Piggery | 11,233.3±274.4 | 15,251.5±143.0 | 11,383.3±563.0 | 14,981.7±561.5 | 11,483.3±563.0 | 15,666.7±726.5 |
| 5. | Backyard poultry | 3,393.3±109.7 | 6,366.7±88.2 | 3,323.3±110.5 | 6,283.3±101.4 | 3,451.7±133.3 | 6,450.0±144.3 |
| 6. | Fishery | 3,983.3±274.4 | 7,016.7±274.4 | 3,643.3±127.1 | 6,420.0±75.7 | 4,116.7±116.7 | 7,146.7±163.3 |
| 7. | Water management | 150.0±5.8 | 2,000. ±80.9 | 150.0±5.8 | 2,000.3±1000.3 | 199.3±13.9 | 2100.3±118.0 |

**Gross Returns of different Component of Farming Systems at Chandel, Churachandpur and Ukhrul under CFS and IIFS**

| S.N. | Components | Gross Return (Rs) | | | | | |
|---|---|---|---|---|---|---|---|
| | | Chandel | | Churachandpur | | Ukhrul | |
| | | CFS | IIFS | CFS | IIFS | CFS | IIFS |
| 1. | Paddy cultivation | 10,466.7±260.3 | 16,666.7±296.3 | 10,233.3±202.8 | 16,213.3±225.8 | 14,400.0±577.4 | 23,070.0±915.2 |
| 2. | Vegetable cultivation | 15,200.0±544.0 | 22,533.7±606.9 | 16,033.7±289.3 | 23,883.1±709.3 | 19,033.3±448.1 | 28,516.3±591.7 |
| 3. | Fruits cultivation | 4,550.0±217.9 | 7,533.3±233.3 | 4,900.0±144.3 | 7933.3±260.3 | 5,500.0±230.9 | 9,076.7±159.2 |
| 4. | Piggery | 26,366.7±721.9 | 46,033.3±1,414 | 24,133.3±638 | 43,080.0±1,109 | 27,466.7±868 | 48,393.3±1,130 |
| 5. | Backyard poultry | 18,966.7±837.3 | 38,400.0±1,078.6 | 19,650.0±361.7 | 36,100.0±1,126.9 | 20,733.3±520.7 | 38,306.7±1,190.2 |
| 6. | Fishery | 24,900.0±838.7 | 37,433.3±1,097.5 | 22,400.0±608.3 | 37,033.3±1,183.7 | 26,366.7±696.0 | 41,133.3±638.6 |
| 7. | Water management | 350.0±6.90 | 5,500.7±155.2 | 320.0±5.01 | 5,000.0±181.5 | 450.0±8.3 | 6,000.3±274.4 |

**Net Returns of different Component of Farming Systems at Chandel, Churachandpur and Ukhrul under CFS and IIFS**

| *S.N.* | *Components* | *Net Return (Rs)* | | | | | |
|---|---|---|---|---|---|---|---|
| | | *Chandel* | | *Churachandpur* | | *Ukhrul* | |
| | | *CFS* | *IIFS* | *CFS* | *IIFS* | *CFS* | *IIFS* |
| | CFS | IIFS | CFS | IIFS | CFS | IIFS | |
| 1. | Paddy cultivation | 6,024.2±347.3 | 10,514.2±214.8 | 5,923.0±326.4 | 10,161.2±331.3 | 8,379.3±407.8 | 14,616.0±774.0 |
| 2. | Vegetable cultivation | 12,557.5±576.7 | 19012.9±635.3 | 13,222.0±348.8 | 20,274.8±687.2 | 15,701.3±376.4 | 24075.5±648.3 |
| 3. | Fruits cultivation | 3,048.3±187.8 | 5,291.7±154.3 | 3,356.5±200.8 | 5,618.3±162.1 | 3,956.7±335.7 | 6,761.7±46.9 |
| 4. | Piggery | 15,133.3±996.0 | 30,781.8±1,271.6 | 12,750.0±104.1 | 28,098.3±1,670.7 | 15,983.3±308.7 | 32,726.7±1,856.9 |
| 5. | Backyard poultry | 15,573.3±727.6 | 32,033.3±990.5 | 16,326.7±472.1 | 29,816.7±1,026.5 | 17,281.7±652.8 | 1,046.1 |
| 6. | Fishery | 20,916.7±1,112 | 30,416.7±1,371.8 | 18,756.7±481.5 | 30,613.3±1,258.6 | 22,250.0±579.5 | 33,986.7±797.9 |
| 7. | Water management | 200.0±15.8 | 3,500.0±216.7 | 170.0±15.8 | 3,000.3±181.9 | 250.7±116.3 | 3,900.0±250.0 |

**Benefit Cost Ratio (BCR) of different Component of Farming Systems at Chandel, Churachandpur and Ukhrul under CFS and IIFS**

| *S.N.* | *Components* | *Benefit : Cost Ratio (Rs)* | | | | | |
|---|---|---|---|---|---|---|---|
| | | *Chandel* | | *Churachandpur* | | *Ukhrul* | |
| | | *CFS* | *IIFS* | *CFS* | *IIFS* | *CFS* | *IIFS* |
| 1. | Paddy cultivation | 2.4±0.1 | 2.7±0.1 | 2.4±0.1 | 2.7±0.1 | 2.4±0.1 | 2.7±0.1 |
| 2. | Vegetable cultivation | 5.75±0.3 | 6.40±0.3 | 5.70±0.3 | 6.62±0.2 | 5.73±0.1 | 6.33±0.3 |
| 3. | Fruits cultivation | 3.0±0.1 | 3.4±0.1 | 3.2±0.2 | 3.4±0.1 | 3.6±0.4 | 3.9±0.1 |
| 4. | Piggery | 2.3±0.1 | 3.0±0.1 | 2.1±0.1 | 2.9±0.2 | 2.4±0.1 | 3.1±0.2 |
| 5. | Backyard poultry | 5.6±0.1 | 6.0±0.1 | 5.9±0.3 | 5.7±0.1 | 6.0±0.4 | 5.9±0.1 |
| 6. | Fishery | 6.4±0.6 | 5.4±0.4 | 6.2±0.1 | 5.8±0.3 | 6.4±0.1 | 5.8±0.2 |
| 7. | Water management | 2.33±0.1 | 2.75±0.1 | 2.13±0.1 | 2.5±0.1 | 2.26±0.2 | 2.87±0.1 |

**Employment Generation Under CFS (A), Employment Generation (B) Employment Generation for Women (C), Employment Generation for Men (D) and Total Employment Generation Based on Group (E) under CFS and IIFS.**

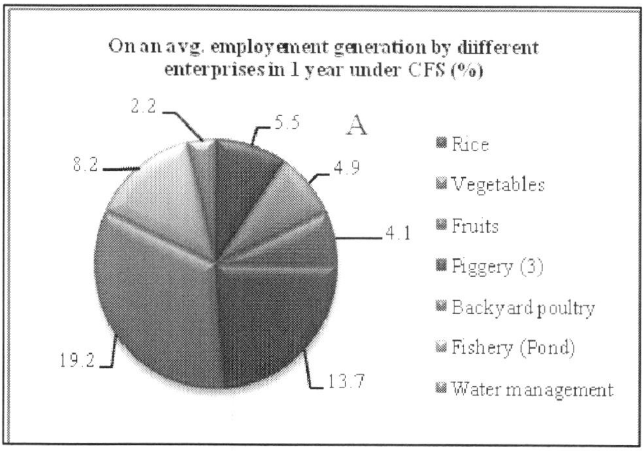


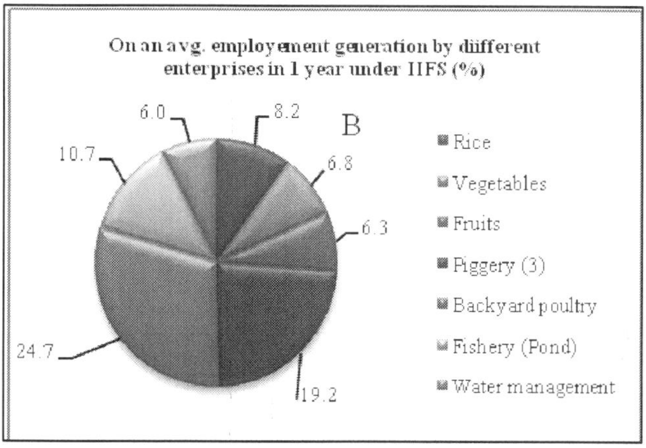


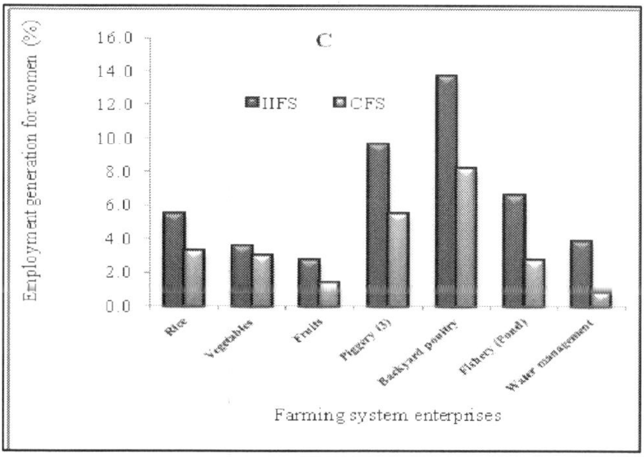


Contd...

***Contd...***

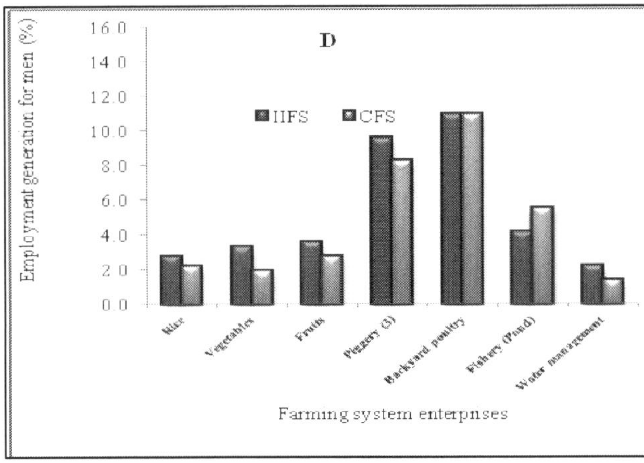


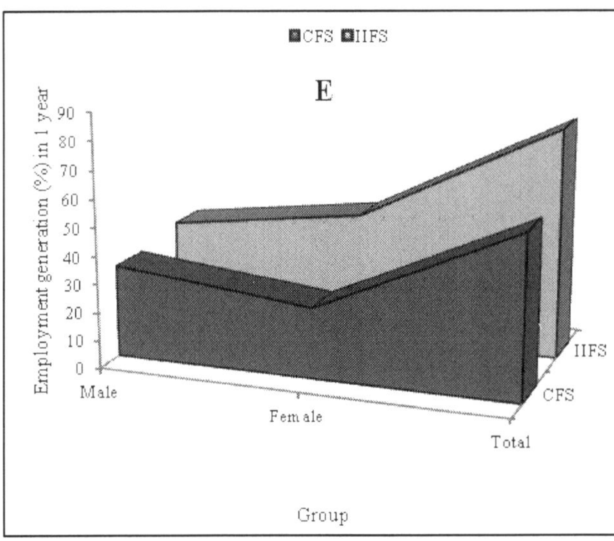


## Case Study II: Analysis of Livestock Feeding Management under Traditional and Improved Farming System among Tribal Farmers of Manipur

In all three districts, commonly piggery, poultry, duckery, goatery, dog rearing and dairy were prevalent among the tribal farmers. The locally available feeds for feeding of livestock were banana stem, Colocasia shoot, paddy straw, Eichhornia and household waste. But under improved farming system, farmers are aware about the feeding management for better health of livestock. Besides locally available feeds, maize grain, broken grain, rice bran, starter feed (Piggery and poultry), pedigree feed (Dog) as well as feed supplements (minerals, mineral mixture, probiotics etc.) are given to the livestock under IFS. During the study, we found that farmers under IFS farmers were vaccinating animals to protect them from various chronic and zonatic diseases. But, under TFS farmers were unaware about the importance of vaccination. During the study period, the multi disciplinary scientific personnel from ICAR Manipur Centre visited the farmers' household on pre-scheduled time and collected necessary information from interviews using structured questionnaires. The interviews were conducted individually as well as in triangulation with the respondents. However, while interviewing in the groups, participant observation method was also employed to observe the farmers' perceptions and recording knowledge on profitability of different livestock feeding under comparative farming systems. Several visits in the rural areas were conducted to collect the information and verification of these practices. The data on cost of different inputs and return from outputs for different systems (TFS and IFS) were also recorded for each farmer. The data collected the outcomes of findings.

**Animal Breeds under different Enterprises**

| *Sl.No.* | *Components* | *TFS* | *IFS* |
|---|---|---|---|
| 1. | Piggery | Local Breed | Cross breeds |
| 2. | Poultry | Local birds | Gramapriya, Vanraja, Improved |
| 3. | Duckery | Local duck | Khaki Campell and Muscovy |
| 4. | Goatery | Undescript breed | Black Bengal |
| 5. | Dairy (Cow, Buffalo) | Local breeds | Cross breed |
| 6. | Dog | Local breeds | Local breeds |

**Price of different Livestock Farming**

| *Particular* | *Price/Unit (Rs)* | *Particular* | *Price/Unit (Rs)* |
|---|---|---|---|
| Piglet | 3000 | Green fodder | 3.0 |
| Pork | 180 | Paddy straw | 2.0 |
| Chicks | 50 | Rice bran | 13.0 |
| Chicken | 160 | Banana stem | 0.50 |
| Local birds | 50 | Colocasia shoot | 0.75 |
| Egg | 10 | Household waste | 2.03 |
| Local birds meat | 200 | Green vegetables (leafy) | 5.0 |
| Goat | 3000 | Maize grain | 12.0 |
| Chewan | 450 | compost | 2.0 |
| Calf (Average) | 5000 | Oil cakes | 20.0 |
| Beef | 200 | Dried stored feed | 4.0 |
| Milk | 30 | Pedigree/Starter feed | 3000-5000 |
| Dog | 1500 | Feed supplement | 4000-9000 |
| Dog meat | 200 | Vaccination | 50 |

## Salient Observations

### Feeding Management

Livestock based farming system is dominant among the tribal farmers of Manipur. Livestock, especially pig and poultry, is a key source of livelihood among the rural families and provides food for consumption, sources of income and capital and socio-cultural needs. Scientific livestock management is a profitable venture because of superior feed conversion ratio, faster growth rate, short generation interval and high proliferation in livestock (Yadav *et al.*, 2013 and Ansari *et al.*, 2013). Imbalanced feeding leads to excess of some nutrients, whilst others remain deficient. This not only reduces meat production and increases costs per kg milk, but also affects various physiological functions including long term animal health, fertility and productivity. To ensure improved productivity, it is necessary to augment and secure feed resources through short and long term planning. In Manipur, Colocasia shoot, Banana stem and Eichhornia are locally available feeds for piggery. Other than this they are also giving household waste and rice bran. In traditional farming system, farmers are solely dependent upon the available feed materials. But, these feeds are not a complete feed. However, under TFS farmers are not giving optimum quantity of feed. On the another hand, improved farming system farmers were aware about the quality livestock feeding. They are also storing feed in the form of hay and silage for lean period. In addition to local feeds, farmers are giving feed in optimum quantity and supplementing them with other quality materials viz; mineral mixture, probiotics and marketed feed. These makes the feed as a balanced feed. The use of agro-industrial by-products (grains, brans and oilcakes) is low and mainly restricted to milch animals and commercial poultry sector. The use of grain for animal feed is less than 5 per cent (except maize), and a lion's share of this goes for poultry feed. Thus crop-residues are by far the most important source of feed and will continue to be so in the foreseeable future. However, their nutritive value is low, thereby, adversely effecting livestock productivity. Several technologies have been developed to improve their nutritive value like, hay and silage making, urea ammoniation and urea molasses treatment or straw etc. The available feeds can be integrated with high quality feeding materials (concentrate) which can act as a complete feed for livestock. Supplementation of concentrates to grazing yaks substantially reduces mortality in winter season when scarcity of feed occurres (Ghosh and Bandyopadhyay 2007).

### Feed Supplementation with Locally Available Feeds

During the study, we found that farmers were not using feed supplements with locally available feeding materials in traditional farming system. Under TFS, almost farmers are not interested in quality feeding of livestock. In Churachandpur and Chandel farmers are not giving any feed supplementation except poultry for some time. But, under improved farming system, farmers giving supplements as mineral mixture, marketed feed, probiotics with feed, oilcakes and supplementing various minerals along with locally available feed materials. Thus, the production and health of livestock was not up to the mark in TFS as compared to IFS. Thomas *et al.*, 2002 also observed that the low productivity of livestock reflects the non adoption of technologies or their uptake has not been sustainable. Adoption of improved technology is low.

The optimum amount and composition of supplementary feeds has been lacking because knowledge of animal nutrition requirements is still deficient. Under this study, we found that farmers are giving 0.5 to 1.5 kg of supplement daily for each animal throughout the dry season. It was estimated that this level of supplementation would enable the animals to maintain the weight. Murugan *et al.* (2009) found that pigs fed with concentrate feed had significantly ($P<0.01$) higher hot carcass weight, dressing percentage and loin eye area, meat-bone ratio and lesser back fat thickness and gut weight than swill feed.

**Quantity of Feed Used under TFS and IFS in Hill Districts of Manipur**

| Sl.No. | Components | Churachandpur (kg/unit/day) | | Chandel (kg/unit/day) | | Ukhrul (kg/unit/day) | |
|---|---|---|---|---|---|---|---|
| | | TFS | IFS | TFS | IFS | TFS | IFS |
| 1. | Piggery | 4.5 | 7.0 | 6.5 | 7.0 | 3.5 | 7.5 |
| 2. | Poultry | 0.44 | 0.55 | 0.45 | 0.55 | 0.35 | 0.65 |
| 3. | Duckery | 0.25 | 0.50 | 0.30 | 0.50 | 0.30 | 0.50 |
| 4. | Goatery | 4.0 | 5.0 | 3.0 | 4.5 | 3.0 | 5.0 |
| 5. | Dairy | 7.0 | 10.5 | 5.5 | 10.5 | 9.0 | 11.5 |
| 6. | Dog | 0.65 | 1.0 | 1.1 | 1.0 | 0.70 | 1.25 |

**Feed supplement (probiotics, marketed feed, minerals and mineral mixture) used under TFS and IFS**

| Sl.No. | Components | Churachandpur (kg/unit/day) | | Chandel (kg/unit/day) | | Ukhrul (kg/unit/day) | |
|---|---|---|---|---|---|---|---|
| | | TFS | IFS | TFS | IFS | TFS | IFS |
| 1. | Piggery | – | + | – | + | ± | + |
| 2. | Poultry | ± | ± | – | + | ± | + |
| 3. | Duckery | – | + | – | + | – | + |
| 4. | Goatery | – | + | – | + | – | + |
| 5. | Dairy | – | + | – | + | – | + |
| 6. | Dog | – | + | – | + | – | + |

–, + and ± represents the feed supplement are not given, feed supplement are given and feed supplement given seldom, respectively

## Vaccination and Medical Care of Livestock

Livestock medical care is equally important to maintain their health. Effective control of animal diseases is a critical element of any strategy that aims at supporting and enhancing livestock dependent livelihoods of rural poor. The livestock mainly suffered from foot and mouth disease, black quarter, Haemorrhagic Septicaemia in animal and influenza virus in birds. The following figures show that Churachandpur, Chandel and Ukhrul district farmers were not aware about the vaccination at birth or

**Feeding Management under TFS and IFS in Hill Districts of Manipur**

| Sl.No. | Components | Churachandpur | | Chandel | | Ukhrul | |
|---|---|---|---|---|---|---|---|
| | | TFS | IFS | TFS | IFS | TFS | IFS |
| 1. | Piggery | a. Household waste<br>b. Rice bran<br>c. Banana stem<br>d. Colocasia shoot | a. Household waste<br>b. Rice bran<br>c. Banana stem<br>d. Colocasia shoot<br>e. Broken grains<br>f. Starter feed<br>g. Feed supplement | a. Banana stem<br>b. Rice bran<br>c. Food waste | a. Household waste<br>b. Rice bran<br>c. Banana stem<br>d. Broken grains<br>e. Maize<br>f. Starter feed<br>g. Feed supplement | a. Banana stem<br>b. Paddy straw<br>c. Colocasia shoot<br>d. Waste food | a. Household waste<br>b. Rice bran<br>c. Banana stem<br>d. Colocasia shoot<br>e. *Eichornia* spp.<br>f. Broken grains<br>g. Starter feed<br>h. Maize<br>i. Feed supplement |
| 2. | Poultry | a. Rice bran<br>b. Green vegetable<br>c. Household waste<br>d. Maize grain<br>e. Broken rice grain<br>f. Rice bran<br>g. Feed supplement | a. Rice bran<br>b. Starter feed<br>c. Household waste<br>d. Maize grain<br>e. Broken rice grain<br>f. Rice bran<br>g. Feed supplement | a. Rice bran<br>b. Household waste<br>c. Starter feed<br>d. Maize grain<br>e. Broken rice grain<br>f. Rice bran<br>g. Feed supplement | a. Rice bran<br>b. Household waste<br>c. Starter feed | a. Rice bran<br>b. Household waste | a. Rice bran<br>b. Household waste |
| 3. | Duckery | a. Household products<br>b. Rice bran<br>c. Feeding in water bodies | a. Household products<br>b. Rice bran<br>c. Feeding in water bodies<br>d. Feed supplement<br>e. *Eichornia* | a. Household products<br>b. Rice bran<br>c. Feeding in water bodies<br>d. *Eichornia* | a. Household products<br>b. Rice bran<br>c. Feeding in water bodies<br>d. Feed supplement<br>e. *Eichornia*<br>f. Colocasia shoot | a. Household products<br>b. Rice bran<br>c. Feeding in water bodies<br>d. *Eichornia* | a. Household products<br>b. Rice bran<br>c. Feeding in water bodies<br>d. Feed supplement<br>e. *Eichornia*<br>f. Colocasia shoot |
| 4. | Goatery | a. Green grass<br>b. Pasture<br>c. Maize grain<br>d. Rice bran<br>e. Feed supplement | a. Green grass<br>b. Pasture<br>c. Maize grain<br>d. Rice bran<br>e. Feed supplement | a. Green grass<br>b. Pasture<br>c. Maize grain<br>d. Rice bran<br>e. Feed supplement | a. Green grass<br>b. Pasture | a. Green grass<br>b. Pasture | a. Green grass<br>b. Pasture |

Contd...

**Contd...**

| Sl.No. | Components | Churachandpur | | Chandel | | Ukhrul | |
|---|---|---|---|---|---|---|---|
| | | TFS | IFS | TFS | IFS | TFS | IFS |
| 5. | Dairy (Cow, Buffalo) | a. Paddy straw<br>b. Green grass<br>c. Pasture | a. Paddy straw<br>b. Green grass<br>c. Pasture<br>d. Oil cakes<br>e. Dried feed (Hay, silage)<br>f. Feed supplement | a. Paddy straw<br>b. Green grass<br>c. Pasture | a. Paddy straw<br>b. Green grass<br>c. Pasture<br>d. Oil cakes<br>e. Dried feed (Hay, silage)<br>f. Feed supplement | a. Paddy straw<br>b. Green grass<br>c. Pasture<br>d. Rice bran | a. Paddy straw<br>b. Green grass<br>c. Pasture<br>d. Oil cakes<br>e. Dried feed (Hay, silage)<br>f. Feed supplement |
| 6. | Dog rearing | a. Household wastage | a. Household wastage<br>b. Pedigree feed for pups<br>c. Feed supplements | a. Household wastage | a. Household wastage<br>b. Pedigree feed for pups<br>c. Feed supplements | a. Household wastage | a. Household wastage<br>b. Pedigree feed for pups<br>c. Feed supplements |

just after birth under TFS. Sometimes farmers got their livestock treated when they suffered from disease. In Ukhrul, farmers were careful about the medical care of livestock under TFS. However, under IFS farmers are more aware and careful about medical treatment and disease occurrence at birth and in adult livestock. Homewood *et al.* (2006) reported that the infection-and-treatment method of vaccination has a major and highly significant impact on survival. Uptake of vaccination is strongly associated with a measure of wealth that includes livestock numbers and economic security.

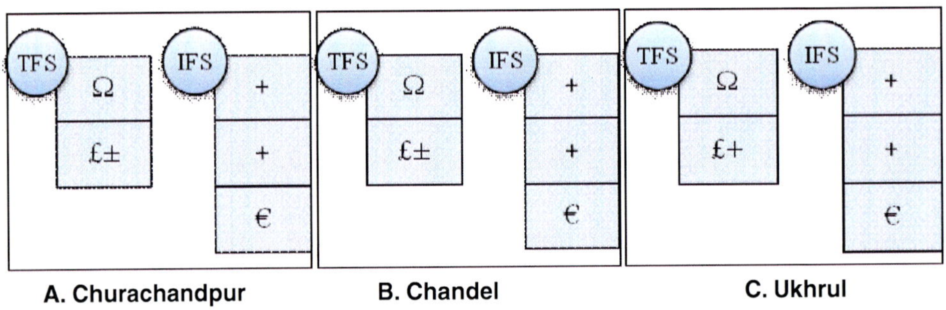


**Vaccination and Medical Care of Livestock**

**Note: Ω represents the seldom vaccination at or just after birth, £± represents some farmers giving vaccination during disease occurrence, £+ represents all farmers are giving vaccination during diseases. + Sign represents the all farmers giving vaccination at or after just birth and during diseases. € represents the regular care of animals to protect from various zonatic diseases.**

## Economics

The current study indicates higher variable cost, gross returns, net returns and benefit cost (B: C) ratio in all the livestock farming of IFS as compared to TFS. Among the districts, Ukhrul has better adoption of farming system either TFS or IFS than other two hill districts of Manipur. The higher variable cost in all the livestock farming of IFS was mainly due to use of quality feed material, improved breeds, proper medication with better feeding management system. The higher gross returns, net returns and B: C ratios were due to more production of economic products in respective components of IFS as compared to TFS. The lower cost of production, gross returns, net returns and B: C ratios in TFS might be due to use of poor feeding management, local breeds with poor medication just for subsistence farming for sustaining the family livelihood. Among the livestock farming, dairy farming followed by piggery farming showed higher variable cost and but higher gross and net returns was observed in piggery followed by backyard poultry. The net return is the main parameter for deciding the adoptability of a farming system.

To ensure improved productivity, it is necessary to augment and secure feed resources through short and long term planning. Balanced livestock feeding management is very important for the livestock health, for their generation as well as for healthy livestock product, which is consumed by humans being mainly as second-cycle produce, in the form of meat, eggs and dairy products. In this way, locally

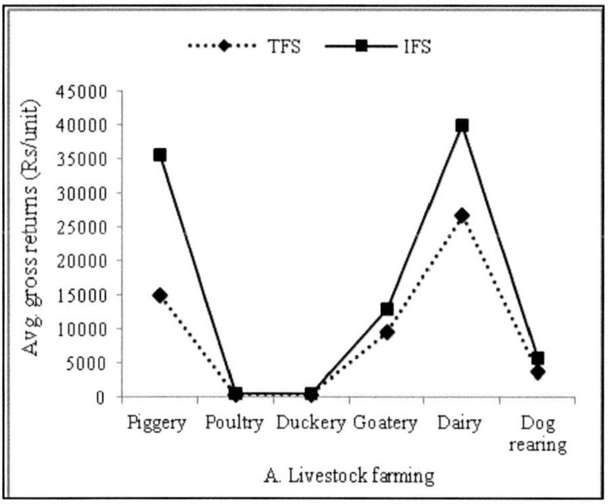


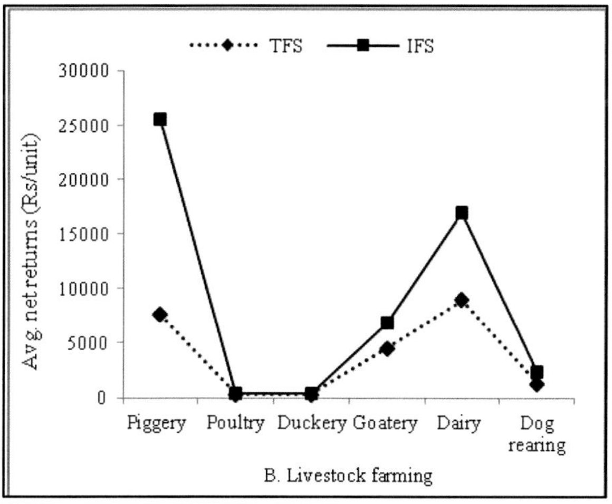


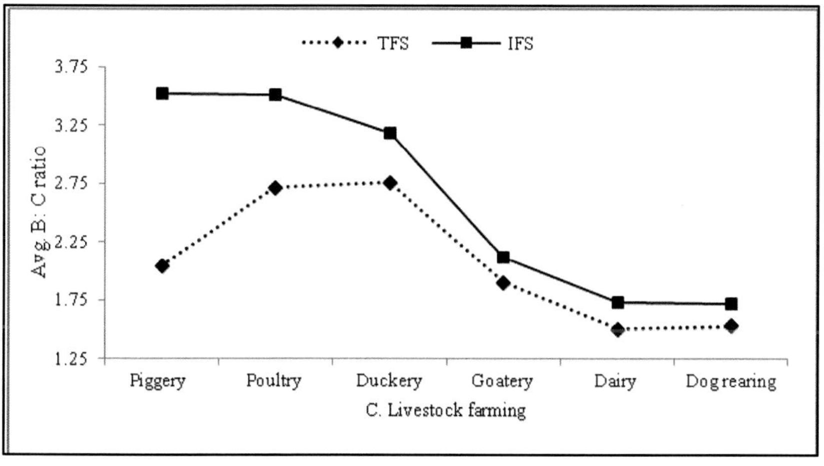


**On an Average Economics (All three districts) under TFS and IFS.**

**Note: on an average including all inputs and output/unit.**

available feed materials should also be supplemented by balanced nutritional feed with proper medication and management to improve farming system. It is of utmost importance to tranform traditional farming system into the improved farming system for betterment of livelihood of the tribal farmers.

# 7
# Success Stories

## Success Story I

| | | |
|---|---|---|
| **Name of the farmer** | : | **Shri A.S. Somi** |
| **Village** | : | **Nungshangkhong Village** |
| **District** | : | **Ukhrul** |
| **Age and Sex** | : | **51 Yrs. (M)** |
| **Land holding** | : | **8.00 ha** |
| **Education** | : | **Graduate** |

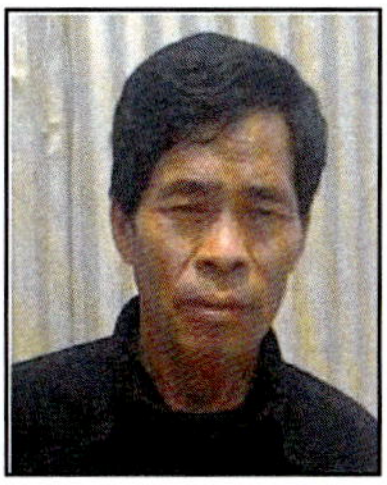

An integrated farming system model was developed in the field of Mr. A. S. Somi in Nungshangkhong village, Ukhrul district of Manipur during 2011–12 to 2014–15 on an area of 6 hectares with adjacent 2 hectares area under natural forest. Mr. Somi was economically very much weaker. He requested ICAR Research Complex for NEH Region, Manipur Centre for technical assistance. ICAR made the technological interventions under Tribal Sub-Plan through high yielding varieties of paddy, groundnut, rapeseed mustard, vegetables, spices, poultry, fishery and jalkund. His IFS models integrated with secondary agriculture like mushroom cultivation, bee keeping and vermicomposting. He has constructed four ponds covering 0.75 hectares which are used for fingerling production. He has also established one circular carp hatchery and producing 1.5 to 2.0 lakhs fingerlings annually. Besides he has constructed one jalkund for providing life saving irrigation to his crops during lean period. Mr. Somi also practice paddy cum fish culture in his paddy field to earn additional profit with minimum investment. Besides, he is also practicing composite fish culture in 2 ponds. In 2011-12, the farmer had no water harvesting unit. After intervention, he could store approximately 75000 litres of water in the Jalkund. In addition to 2 ha area under under field crops (paddy var. RC Maniphou-10 and groundnut var. ICGS-76) he has developed 3.5 ha area under horticulture (cabbage var. Green Hero, cauliflower var. White Flash, brinjal var. Pusa Kranti, french bean var. Contender, pea var. Arkel, tomato var. RC Manikhamenashinba-I, ginger var. Nadia, local varieties of king chilli and broad bean). The fruit crops include kiwifruit (Allison and Hayward), khasi mandarin and kachai lemon. After paddy Mr. Somi grows rapeseed mustard (M-27) under zero tillage. He integrated bee-keeping (5 boxes) with the rapeseed mustard and agro-forestry component. In livestock component he has 5 goats (Assam Local) and 520 poultry birds (Vanaraja). Besides, he also established one vermicomposting unit and one mushroom production unit. The crop residues is used for composting whereas the poultry dropping is used for feeding the fishes. The paddy straw he is using for mushroom production and the spent mushroom substrate is again used for composting. He successfully harvested 4.20 t/ha paddy after intervention as compared to 2.85 t/ha before intervention. He has grown groundnut (ICGS-76) for the first time in his village which gave 2.30 t/ha dry pod yield. Vegetable yield also considerably increased after intervention which gave 1.15 lakh gross returns. In poultry, Vanraja is performing well and he is getting 145 eggs/annum/bird. He also started selling goat kid @ Rs. 7000/-. Fishery is the most profitable component in his IFS where he is earning more than 3.00 lakhs annually. From mushroom and apiculture, he earns around Rs. 50000.00 per annum. From 6 ha land on an average he earned the total net returns Rs 3,95,000.00/year under IFS as compared to Rs 65000.00/year in 2011-12 under traditional farming system. Now, he has become a role model for not only in his village but also for the entire tribal community in the district. These interventions made a positive impact on the utilization of scarce resources under fragile hilly ecosystem benefitting tribal farmers at large scale.

## Success Story II

| | | |
|---|---|---|
| **Name of the farmer** | : | **Shri R.D. Peter** |
| **Village** | : | **Purul Akutpa Village** |
| **District** | : | **Senapati** |
| **Age and Sex** | : | **37 Yrs. (M)** |
| **Land holding** | : | **13.00 ha** |
| **Education** | : | **H.S. (10+2)** |

Mr. R. D. Peter, S/o R. Dakho from Purul Akutpa Village in Senapati district has come from very poor family. Due to financial problem he could not complete his graduation and engaged in farming activities. Due to lack of knowledge about improved farming practices he could not earn much from his land. He approached

ICAR Research Complex for NEH Region, Manipur Centre and requested to provide proper guidance. ICAR team visited his farm and found that the area is suitable for developing integrated farming system. He has been trained in various farm technologies. Under the financial assistance from Tibal Sub-Plan, he developed a sustainable model of integrated farming system. Based on sloping agricultural land technology, the integrated farm has an area of 13 ha, in which different agro-forestry species have been grown at the hill top (10 ha), followed by horticultural crops, agricultural crops, animal and fishery component (3 ha) in descending order of elevation. The unique character Mr. Peter has incorporated into his farm is the high value enterprises. His integrated farm now has 8 ponds (with polyethylene lining) covering 1 ha area. He is using the ponds mainly for production of fish fingerlings. Every year he is producing 2.5 to 3.0 lakh fish fries and fingerlings and selling to his fellow farmers. He also established one circular carp hatchery. Besides, Mr. Peter successfully adopted paddy cum fish culture and producing 150-170 kgs of fish/ha almost without any extra investment. He has also established one poultry and one piggery unit with recommended isolation distance. The fodder plants in the agro-forestry system and crop residue from the farm are fed to the animals. Instead of local strain he started rearing Vanaraja poultry birds and emphasized on selling chicks to increase his profits. Annually he sells more than 6000 eggs and 2500 chicks. He is using solar power to maintain temperature inside the poultry unit during night in winter months. Horticulture is one of the major component in his farm. Mr. Peter has kiwifruit plantation where he grows broccoli and leek as intercrop. Besides, he also has one constructed polyhouse where he is growing king chilli. He also grew papaya and chow chow on a small scale. The availability of different agri-inputs in the area is a problem, therefore, he is raising the crops almost organically with the use of vermicompost etc. The crop residue and animal waste is being utilized to make manure and vermicompost to fertilize the crops. Tree bean, wild apple, gooseberry and *Prunus nepalensis* are the major crops under agro-forestry. Every year he sells more than 1500 pods of tree bean and earn Rs. 16200/- annually; whereas *Prunus nepalensis* fruits are utilized for making various value added products. Recently Mr. Peter has obtained FSSAI Food Licence for his small fruit processing unit. He is producing 500-600 litres beverages from various indigenous fruits with net profit of Rs. 25800/-. Both the forest products ensure a huge profit for Mr. Peter almost without any investment. In addition, he has also integrated the horticultural component with bee keeping and annually he harvests around 20 litres of honey from 2 bee boxes and earns a net profit of Rs. 22000/-. Recently he has introduced grapes into his horticulture component. For water conservation, Mr. Peter has not only constructed eight ponds but also made three Jalkunds on the upper side of the hills. The conserved water in Jalkund is used for providing life saving irrigation to the crops during lean period. He is also using drip irrigation for judicious use of water. He also practices mulching for minimizing the soil erosion. Presently he is earning Rs. 5.21 lakhs annually with a B:C ratio of 7.78. With his dedication and hard work, Mr. Peter has not only sustain his livelihood, but also become an inspiration for other farmers.

## Success Story III

| | | |
|---|---|---|
| **Name of the farmer** | : | **Shri Henkpao** |
| **Village** | : | **T. Champhai Village** |
| **District** | : | **Churachandpur** |
| **Age and Sex** | : | **47 Yrs. (M)** |
| **Land holding** | : | **4.00 ha** |
| **Education** | : | **Matric** |

Considering the physical, social and economic limitations of the Manipur, an integrated farming system model was developed in the field of Henkpao (tribal farmers) in T. Champhai village, Churachandpur district of Manipur during 2011–12 to 2014–15. Mr. Henkpao is one such beneficiary who had shown enthusiasm to accept and adopt the technology. The model comprises 4 ha of fenced area in the vicinity of the tribal settlement integrated with seven components. With the assistance of TSP, technological interventions was made through high yielding variety Paddy (RCM-9), Maize (Pusa composite-3), groundnut (ICGS-76), pea (Azad pea-1), vegetables [cabbage (Rareball) and cauliflower (Early Hemlata)], fruits (50 tree bean, 50 kachai lemon and 50 orange), 6 piglets (Cross bred Hampshire), 125 poultry (Gramapriya), fishery (fingerlings of Common carp, grass carp and Catla), agri-polythene sheets of 250 micron meant for Jalkunds and vermicomposting unit for waste recycling. The Capacity building programme on Integrated Farming System was conducted to the villagers in whom the beneficiary farmer also participated. During the implementation process, initially the farmers were skeptical about the model but in-due course crops, livestock and vegetable in their backyard and the fast growing healthy chicks and pigs changed their attitude. For the first time in Churachandpur district of Manipur, the farmers harvested a bumper crop of vegetables in their field. He successfully harvested higher paddy under IFS 4.80 t/ha after intervention (AI) as compared to 3.25 t/ha rice yield before intervention (BI) in subsistence farming system (SFS). Farmer had grown first time groundnut (ICGS-76) which gave 2.40 t/ha dry pod yield. Maize cultivar Pusa composite-3, produced 3.25 t/ha as compared to local maize cultivar (Chakhaocujak (hill)) yield 1.70 t/ha. Farmer's vegetable yield considerably increased after intervention which gave 1.50 lakh net returns from cabbage and cauliflower. In the 2nd year, the 15 number of piglets increased including with 6 pigs. In poultry farming, Gramapriya is performing well and farmers were getting 40-45 eggs/day. He adopted composite fish production in which grass carp was surviving on middle and upper layers of water and common carp in lower layers of water. In 2011-12, the farmer had no water harvesting unit. In 2013-14 and 2014-15, he could store 30,000.00 and 55000 litres of water in the Jalkund respectively. The net farm income was realized by farmer who maintained *"crop-livestock-fish-vegetable-fruits-water harvesting" integration on their field*. In improved practice, from 4 ha land on an average/year he earned the total net returns Rs 3,63,500/[(Paddy cultivation (2 ha)= Rs 82000, Groundnut production (0.5 ha)= Rs. 38000, Maize production as green cob (0.5 ha)= Rs. 23000, Vegetable production in rabi season (1 ha, Cabbage and Radish)=Rs. 150000, Fruit production (Fruiting not started)=Nil, Piggery= Rs. 37000, Poultry= Rs.

16000, Fishery= Rs. 17500] as compared to 105000 in 2011-12 under SFS. Now, he has become a role model for not only in his village but also for the entire tribal community of this district. Now, farmers from other villages were motivated by the success of these farmers and most of the village headmen are requesting to implement such model in their villages. These interventions made a positive impact on the utilization of scarce resources under fragile hilly ecosystem benefitting tribal farmers at large.

## Success Story IV

| | | |
|---|---|---|
| **Name of the farmer** | : | **Shri H. B. Starson** |
| **Village** | : | **Chandel Khullen** |
| **District** | : | **Chandel** |
| **Age and Sex** | : | **31 Yrs. (M)** |
| **Land holding** | : | **2.21 ha** |
| **Education** | : | **Graduation** |

The IFS model was developed in the field of H. B. Starson (tribal farmers) in Chandel Khullen village, Chandel district of Manipur during 2011–12 to 2014–15. A model of integrated farming system was developed on farmer's field. Base line survey on socio-economic as well as their agriculture farming has been done in 2011-12. In 2014-15, survey was done to know the socio economic status of the farmers. The holding size of the farmer was 1.33 ha, 2.06 ha, 2.21 ha in 2011-12 to 2012-13, 2013-14 and 2014-15, respectively due to technological interventions like terracing and thereby fallow land have been put for cultivation of crops. Training and critical inputs were given to the farmers for adoption of scientific management practices in the integrated farming system.

He successfully harvested higher paddy under IFS 4.79 t/ha after intervention (AI) as compared to 3.5 t/ha rice yield before intervention (BI) in subsistence farming system (SFS). However, the yield of rice, pea and rapeseed mustard increased by 36.85, 106 and 90 per cent respectively AI as compared to 2011-12. Farmer's vegetable yield considerably increased after intervention and got 103 and 54 per cent more cabbage and onion yield due to adoption of improved cultivation methods and use of quality seeds and better management practices under integrated farming system. After intervention, number of piglets increased from 2 to 22. However, it increased the pork production by about 900 per cent as compared to 2011-12. An earlier farmer was rearing local strains of poultry, but under the farming systems approach, he started rearing of Gramapriya and Vanaraja. After intervention, 777 and 566 per cent more egg and chicken production, respectively was increased as compared to 2011-12.

In 2011-12, the farmer was not having any farm pond. But, after adoption of integrated farming system, he dug two ponds in 2012-13 and one more ponds in 2013-14. In 2013-14, he started fish breeding programme and got very good remunerative returns. He started fish production and produced 900 kg along with 20,000 numbers of fingerlings and it increased to 2250 kg fish with 42000 numbers of fingerlings in 2014-15. In kharif season, rain water was sufficient for agricultural crops. But, in rabi season farmers were facing acute shortage of water. In 2011-12, the farmer had no water harvesting unit. In 2012-13, he stored 65,000.00 litres of water in the pond and the water storage increased to 1, 20,000.00 litres in 2014-15 due to introduction of more number of rain water harvesting units as Jalkund. The net farm income was realized by farmer who maintained ***crop-livestock-fish-vegetable-fruits-water harvesting integration on their field***. In improved practice, from 2.21 ha land on an average/year he earned the total net returns **Rs 2, 93,500**/as compared to **66267/ in 2011-12** under SFS. Now, he has become a role model for not only in his village but also for the entire tribal community of this district. Now, farmers from other villages were motivated by the success of these farmers and most of the village headmen are requesting to implement such model in their villages.

## Success Story V

| | | |
|---|---|---|
| **Name of the farmer** | : | **Shri V. Tuime lolly** |
| **Village** | : | **Kachai** |
| **District** | : | **Ukhrul** |
| **Age and Sex** | : | **53 Yrs. (M)** |
| **Land holding** | : | **2.21 ha** |
| **Education** | : | **Graduate** |

The Kachai village is one of the most remote villages in Manipur and is located at 25°14′29″N latitude and 94°16′22″E longitude which is extreme west in Ukhrul District, Manipur, forming a boundary with Senapati district. The village is about 140 kms away from Imphal (State Capital) and 50 kms from Ukhrul (District Headquarter). The Kachai village is inhabitant by Tangkhul Naga Community comprising of 360 household and population of 2600. The region is characterized by marginality, fragility and inaccessibility. The rugged terrain hampers the delivery of various services to the farmers. The cost of cultivation is also high due to undulating lay of the land. Ninety five per cent of the soils of this region are acidic in nature and suffer from micronutrient deficiency. During winter months, the regions suffer from severe water scarcity. Farm mechanization is almost non-existent in the region. Good roads and communication network are keys to development. But, the transport network in the region is in a very poor shape. Due to remote locality and inaccessibility, the agri-input delivery and agri-service delivery system in the region is almost absent. Moreover, majority of the farming community is small and marginal. They do not have access to easy credit from the banking institutions, as a result of which they are compelled to continue with small scale household production system.

Mr. V Tuime Lolly is a dedicated, sincere and hardworking farmer of Kachai village. With a strong determination for ensuring a sustainable livelihood, Mr. V. Tuime Lolly sought help from ICAR Research Complex for NEH Region, Manipur Centre. The ICAR team surveyed the area and undertook the programme by conducting residential training and demonstration programme for the farmers. The team also undertook one activity to develop citrus based community farming system with a group of farmer led by Mr. Tuime Lolly at Kachai village. An area of 20 hectares was selected for the development of community farming system. The land was sloppy and the soil was acidic in nature (pH 4.8). Hence, land development activity was undertaken and about 11 hectares sloppy was converted to bench terraces and half moon terraces. Eight hectares of area on the hilltop was developed as agroforestry block with tree bean, fodder trees and mulberry. The remaining area was divided in to different blocks (citrus plantation 6 ha, other fruits block 1 ha, field crops 2 ha and vegetable crops 1 ha). A pond of 2500 sq m area (50 m X 50 m) was also excavated for rearing fishes and utilizing the water to provide life saving irrigation during lean period. About 5000 fish fingerlings of common carp were released in to the pond for composite fish culture. In the livestock component, a dairy unit with 56 cattle and a piggery unit of 10 pigs were established. Based on sloping agricultural land technology (SALT), agro-forestry block was established at the hill top, followed by horticultural crops, agricultural crops and animal and fishery component in descending order of elevation. In addition, one jalkund, one small sericulture unit, vermicomposting unit and one fruit processing unit were also developed. Quality seed and planting material of improved crop verities were provided to the farmers along with other critical inputs.

French bean and garden pea were intercropped in citrus plantation to enhance the soil fertility. Sweet potato was planted on slopes as cover crop. For the first time in the area, biological control method (*Trichoderma* and *Pseudomonas*) was introduced for plant protection. Liming was done to reduce the soil acidity. The availability of different agri-inputs in the area is a problem, therefore, crop residue and animal waste were utilized to make manure and vermicompost to fertilize the crops. Mulberry from agroforestry block was used as a raw material for the sericulture unit. The fodder plants in the agro-forestry system along with maize and soybean plants (after harvest) were used supply the fodder to the animals. The banana pseudostem and sweet potato leaves were used as pig feed. *In-situ* soil and moisture conservation measures were adopted using mulching with locally available material. Catch pits and contour trenches were constructed to recharge the ground water and to reduce the runoff and soil erosion, respectively. Each and every component of the farming system were linked in an integrated way. All these measures have resulted in improved productivity of the system and quality of the produce.

# References


19th Livestock Census - 2012 All India Report, Govt. of India Department of Animal Husbandry, Dairying and Fisheries, Ministry of Agriculture and Farmers Welfare, Govt. of India.

Ansari M A, Prakash N, Baishya L K, Punitha P, Sharma P K, Yadav J., and Kabui G P and Kl Levis Ch. 2014. Integrated Farming System: an ideal approach for developing more economically and environmentally sustainable farming systems for the Eastern Himalayan Region. *Indian Journal of Agricultural Sciences* 84 (3): 356–62.

Ansari M A, Prakash N, Baishya L K, Punitha P, Yadav J S, Sharma P K, Sailo Blessa and Ansari M H. 2013. Comparative study on conventional and improved integrated farming systems for sustainable production, income generation and employment opportunity among the tribal farmers in hilly Regions of Manipur. Indian Journal of Agricultural Sciences 83 (7): 765–72.

Ansari M A, Prakash N, Punitha P., Roy S S, Sharma S K, Panwar A S and Ngachan S V.2015. On farm participatory research on farming system: enhanced the livelihood and employment generation among tribal farmers of North East India. Abstracts of International conference on natural resource management for food security and rural livelihoods during 10 13 February, 2015, New Delhi. Pp: 500.

Area and Production of Major Agricultural Crops 2013-14. Department of Agriculture, Government of Manipur.

Area and Production of Major Horticultural Crops 2013-14. Department of Horticulture and Soil Conservation, Government of Manipur.

Basic Animal Husbandry and Fisheries Statistics 2014, Department of Animal Husbandry Dairying and Fisheries, Ministry of Agriculture, Government of India.

Behera U K and Mahapatra I C. 1999. Income and employment generation of small and marginal farmers through integrated farming systems. *Indian Journal of Agronomy*. 44(3): 431-39.

Behera U K and Sharma A R. 2007. Modern Concepts of Agriculture. Division of Agronomy, Indian Agricultural Research Institute, New Delhi-110012.

Behera U K, Jha K P and Mahapatra I C. 2004. Integrated management of available resources of the small and marginal farmers for generation of income and employment in eastern India. *Crop Research* 27(1): 83-89.

CARDI. 2010. *A Manual on Integrated Farming Systems (IFS)*. 57 pages, Caribbean Agricultural Research and Development Institute. Ministry of Economic Development, Belize.

Chan, G.L., 2006. Integrated Farming System. What Does Integrated Farming System Do? http://www. scizerinm.org/chanarticle.htm.

Christen Olaf and O'Halloranetholtz Zita. 2002. *Indicators for a Sustainable Development in Agriculture*. 20 pages, European Initiative for Sustainable Development in Agriculture (EISA), Germany.

Dent J B.1990. *Systems theory applied to agriculture and the food chain*. Elsevier, Amsterdam.

Economic Survey Manipur 2013-14. Directorate of Economics and Statistics, Government of Manipur.

FAO. 2001. *Farming Systems and Poverty: Improving Farmers' livelihoods in a changing World*. Food and Agriculture organization of the United Nations, Rome. pp 412.

Fisheries Statistics 2013-14. Department of Fisheries, Govt. of Manipur.

Ghosh M K and Bandyopadhyay S. 2007. Present nutritional status and future strategies for sustainable yak husbandry in India. *Indian Farming* **56**: 3-8.

Homewood Katherine, Trench Pippa, Randall Sara, Lynen Godelieve, Bishop Beth. 2006. Livestock health and socio-economic impacts of a veterinary intervention in Maasailand: Infection-and-treatment vaccine against East Coast fever. *Agricultural Systems* **89**(2–3): 248-271.

Igbinnosa, I. and Okporie, 2007. Preparation of Training Tookit for Sustainable Farming System. Consultant Draft Report-Second National FADAMA Development-Critical Ecosystem Management Project, 2007.

Jamir Tongdi and De Uday Shankar. 2013. Trend in GHG Emissions from Northeast and West Coast Regions of India. *Environmental Research, Engineering and Management*. 1(63): 37-47.

Jha D. 2003. An overview of farming systems research in India. *Annals of Agricultural Research* 24(4):695-706.

Kumar S and Jain D K. 2005. Are linkages between crops and livestock important for the sustainability of the farming system? *Asian Economic Review* 47(1):90-101.

Lightfoot C. 1990. Integration of aquaculture and agriculture: a route to sustainable farming systems. *Naga, The ICLARM Quarterly* 13(1):9-12.

Mahapatra I C. 1992. Farming systems research challenges and opportunities. *Eastern Indian Farming System Research and Extension, Newsletter* 6(4):3-10.

Mahapatra I C. 1994. Farming system research – A key to sustainable agriculture. *Fertilizer News*, 39(11):13-25.

Manipur State Action Plan on Climate Change 2013. Directorate of Environment, Government of Manipur.

MBRLC. 1988. *A Manual on How to Farm your Hilly Land without Losing your Soil.* Mindanao Baptist Rural Life Center, Davao del Sur, Philippines.

Murugan M, Mathew Joseph, Sivakumar T and Gnanaraj P Tensingh. 2009. Effect of different feeding system on the performance of crossbred pigs. *Indian Journal of Animal Sciences* **79**(10): 1058–1060.

Othman, K., 2006. Integrated Farming System and Multifunctionality of Agriculture in Malaysia. http://www.actahort.org/books/655/655-36.htm.

P. Punitha, N. Prakash, **M. A. Ansari** and P. K. Sharma. 2012. Training integral PRA to extension personnel: Extension experience. 6th National seminar on "Emerging Challenges and Paradigms for Sustainable Agri Rural Development" held at Shimla during 18-20 December, 2012. Pp: 193.

Prakash N, Ansari M A, Roy S. S., Punitha P., Sharma S K and Ngachan S V. 2015. Integrated farming system: opportunities for food security and rural livelihood development in North East India- Case Study. Abstracts of International conference on natural resource management for food security and rural livelihoods during 10-13 February, 2015, New Delhi. Pp: 508.

Prakash N, Roy S S, Sharma P K and Ngachan S V. 2012. *Developing the Potential of Underutilized Horticultural Crops of Hill Regions.* ISBN 81-7019-473-3 (India), ISBN 1-55528-330-6 (USA) 608 Pages, (Prakash N, Roy S S, Sharma P K and Ngachan S V Ed.). Todays and Tomorrows Printers and Publishers, New Delhi.

Pretty J. 1995. Regenerating Agriculture: Policies and Practice for Sustainability and Self-Reliance. Earthscan Publications, London.

Rana S S and Chopra Pankaj. *Integrated Farming System.* 90 pages, Department of Agronomy, College of Agriculture, CSK HPKV, Palampur, Himachal Pradesh.

Rana S S and Rana M C. 2013. *Modern Concepts in Crop Production.* 144 pages, Department of Agronomy, College of Agriculture, CSK HPKV, Palampur, Himachal Pradesh.

Rangaswamy A, Venkitaswamy R, Purushothaman and Palaniappan SP. 1996. Rice-Poultry-Fish-Mushroom integrated farming systems for lowlands of Tamil Nadu. *Indian Journal of Agronomy*, 41(3): 344-48.

Ravisankar, N., Ganeshkumar B., Ghoshal Chaudhuri S., Raja R., Srivastava R.C., Singh D.R., Kundu A. and Medhi R.P., 2006. Development of Integrated Farming System (IFS) Models under different Resource Conditions in Humid Tropics of Bay Islands.http://cari.res.in/Naturalper cent20rm ach/Developmentper cent20ofper cent 20IFS.htm.

Roy S S, Prakash N, Sharma P K, Singh I M and Ngachan S V. 2012. *Principles and Concepts of Organic Farming*. 136 pages, ICAR Research Complex for NEH Region, Umroi Road, Umiam, Meghalaya.

Roy S S, Sharma S K, Ansari M A, Banerjee A, Deshmukh N A, Prakash N and Ngachan S V. 2014. Integrated Farming System for Sustainable Agriculture. *Integrated Farm Management*, pp 1-24. Gupta Asha and Saxena Vijay Laxmi Eds. Aaviskar Publishers, Distributors, Jaipur, India.

Shaner W W, Philipp P F and Schmehl W R. 1982. *Farming Systems Research and Development: Guide for Developing Countries*. West View Press, Colorado, USA.

Singh G. 2005. Farming systems options for sustainability of natural resources. Proceedings of Symposium on Alternative Farming Systems: Enhanced income and employment generation options for small and marginal farmers (Singh A K, Gangwar B and Sharma S K Eds.). pp 57-64. 16-18 September, 2004. Project Directorate for Cropping Systems Research, Modipuram.

Singh H P, Sharma K D, Subba Reddy G and Sharma K L. 2004. Dryland Agriculture in India. Challenges and Strategies for Dryland Agriculture. CSSA Special publication No. 32. Crop Science Society of America and American society of Agronomy 671S., USA p. 67-92.

Singh Kalyan, Bohra J S, Singh Y and Singh J P. 2006. Development of farming system models for the north-eastern plain zone of Uttar Pradesh. *Indian Farming* 56 (2): 5-11.

Singh R P and Subba Reddy G. 1986. Research on drought problems in arid and semi-arid tropics. ICRISAT, Patancheru, India.

Tokrishna, R., 2006. Integrated Livestock-Fish Farming Systems in Thailand. http://www.fao.org/docrep/004/ac155e/AC155E13.html.

Vision 2030, ICAR Research Complex for NEH Region, Umiam, Meghalaya.

Vision 2030, Project Directorate of Farming System Research, Modipuram, Meerut, Uttar Pradesh.

Yadav Gulab Singh, Debnath Chandan, Datta M, Ngachan S V, Yadav J S and Babu Subhash. 2013. Comparative evaluation of traditional and improved farming practices in Tripura. *Indian Journal of Agricultural Sciences* **83** (3): 310–4.

Web-link

http://agritech.tnau.ac.in/fishery/fish_ifs.html

http://ec.europa.eu/environment/archives/ppps/pdf/ilubrochure.pdf

http://www.authorstream.com/Presentation/nasirdar-476664-integrated-farming-systems/

http://www.crida.in/DRM2-Winter%20School/GSR-VM.pdf

http://www.fao.org/farmingsystems/description_en.htm

http://www.hillagric.ac.in/edu/coa/agronomy/lect/agron-4711/Lecture%203%20System%20approach.pdf

http://www.hillagric.ac.in/edu/coa/agronomy/lect/agron-4711/Lecture%201%20Farming%20system%20scope%20importance%20and%20concept.pdf

# Index

*F*



*G*